FAILURES IN CONCRETE STRUCTURES

Case Studies in Reinforced
and Prestressed Concrete

FAILURES IN CONCRETE STRUCTURES

Case Studies in Reinforced *and* Prestressed Concrete

Robin Whittle

CRC Press
Taylor & Francis Group
Boca Raton London New York

CRC Press is an imprint of the
Taylor & Francis Group, an **informa** business

A SPON PRESS BOOK

CRC Press
Taylor & Francis Group
6000 Broken Sound Parkway NW, Suite 300
Boca Raton, FL 33487-2742

First issued in paperback 2018

© 2013 by Robin Whittle
CRC Press is an imprint of Taylor & Francis Group, an Informa business

No claim to original U.S. Government works

ISBN 13: 978-1-138-07423-1 (pbk)
ISBN 13: 978-0-415-56701-5 (hbk)

Visit the Taylor & Francis Web site at
http://www.taylorandfrancis.com

and the CRC Press Web site at
http://www.crcpress.com

Contents

Foreword

Errare humanum est. ... We structural engineers are human and so have made a number of errors over the years resulting in narrow escapes, badly performing structures, and even fatal collapses. But as Seneca continues ... *sed perseverare diabolicum,* we must not repeat our errors.

To avoid this means that we must learn from our past mistakes; we must know what went wrong and why. Some of the lessons from our past errors get embodied in clauses in codes of practice, but many do not, and the collective memory of the profession tends to fade as the generation of engineers who learnt from the mishaps and catastrophes retires.

Past books on the subject of structural failures tended to deal with the general causes of failures and methods of investigation, illustrated with the more spectacular examples. However, details of some failures that have not made the headlines, but nevertheless hold important lessons, are hard to find or may not even be in the public domain.

In the past, Robin Whittle and I worked together at Arup R&D on a variety of problems of concrete structures. Some of these arose from failures, and others were encountered when forestalling undesirable outcomes of the enthusiasm—untempered by experience—of some of our younger colleagues.

Robin was also in close contact with researchers at the now sadly defunct Cement & Concrete Association, the Polytechnic of Central London, and the universities of Leeds, Durham, and Birmingham, and so was privy to much of the background for the initial draft and subsequent revisions of CP110.

Nowadays, a preoccupation with the ever-multiplying minutiae of codes, whether Euro Community or National, can blind designers to the imperatives of first principles. New patterns of procurement and site management also widen the communication gap between design and execution, and exert pressures to adopt shortcuts that sometimes have unforeseen consequences.

With his background, Robin is well placed to present a selection of case studies that have lessons for all of us. This is a book that should be read by those structural engineers who wish to broaden their knowledge by learning from some of the experiences of the last 50 years and also for those who would like to refresh their memories.

Poul Beckmann, F.I. Struct. E., M.I.C.E., M.I.D.A., Hon. F.R.I.B.A.

Acknowledgements

The author would particularly like to thank Poul Beckmann for his contribution to this book. He has provided the descriptions of several of the incidents and his comments on the rest of the text have been greatly valued.

The contributions from the following people are also gratefully acknowledged.

Ian Feltham	Steve McKechnie
Terry Bell	Howard Taylor
Alf Perry	Robert Vollum
Geoff Peattie	Richard Scott
Tony Stevens	Andrew Beeby
Tony Jones	Rob Kinch
Chris Kenyon	Gill Brazier
John Hirst	Andrew Fraser
Bryn Bird	

Introduction

This book is a personal selection of incidents that have occurred related to reinforced and prestressed concrete structures. Not all have led to failures and some of the mistakes were discovered at the design stage. Each incident required some form of remedial action to ensure safety of the structure. Some of the incidents were caused by mistakes in design or construction or both. Some involved collapse of part of the structure, but in such cases the cause was from more than one unrelated mistake or problem. A few of the errors and incidents were caused by deliberate intent.

Chapters 1 to 11 describe specific incidents such as structural misunderstanding, extrapolation of codes of practice, detailing, poor construction, and other factors. When a particular incident involved more than one of these causes, it is described in the most relevant section. Chapters 12 and 13 discuss issues related to procurement and research and development.

Care has been taken not to name the particular projects in which the incidents occurred, and the intention in providing the information is to ensure that such mistakes can be understood and avoided in the future.

Some of the problems were discovered in association with requests for support about a different topic. In trying to discover the details of the problem it became clear that other more serious issues were at stake. This begs the question, how many unresolved problems and mistakes are out there that have not seen the light of day? It is fortunate that most reinforced and prestressed concrete structures are indeterminate and allow alternative load paths to form and prevent failures that were not foreseen in the design.

There is a worrying trend in both the design and construction of building structures. Material strengths of both concrete and steel have continued to increase for the past century. At the same time, the overall safety factor used in design has been reducing. This trend is largely caused by the pressure to reduce costs. Inevitably the time will come when the effect of this trend will mean that the errors made in design and construction will not be absorbed by the global safety factor or the ability of a structure to find alternative paths for the loads. This increase in risk is not helped by the changes in the form of contracts and contractual procedure. Health and safety clauses do not provide, in the author's opinion, the necessary safety nets.

In order to improve the quality of all aspects of design and construction, it is essential to involve people who know what they are doing. More training is required at the operational levels of both design and construction. More research and scientific development are required to allow better understanding of the materials and their uses. But the more sophisticated the materials and tools become, the more intelligence is required for their control.

Chapter 1

Failures due to Design Errors

Many failures, when investigated, have been found to arise from a combination of causes. The traditional design sequence starts with the sizing of members. These are determined from the loading (permanent and variable actions) with reference to bending moments and, for beams, shear forces. The reinforcement is then calculated to cater for these forces. Much of the reinforcement is detailed only after completion of the contract documents. Later, if problems are found in fitting the required reinforcement into an element or joint, it is difficult to change the size of section on which the architect and services engineers agreed. Many such problems could have been avoided by producing sketches early on to show how the joint details could work before sizes were finalised.

Computer software is commonly used to provide design information. This cannot always be tailored to suit the problem exactly. All too often a designer fails to ensure, with manual checks, that the software provides a reasonable solution for the particular design. The effects of creep, shrinkage, and temperature are often not considered by the software, and manual calculations are required to check whether effects are significant.

Robustness has become an important consideration in design and this is closely linked to the detailing of joints. The elements of an in situ structure are mechanically connected together by normal detailing and this should provide sufficient robustness. However, for hybrid structures containing a mixture of precast and in situ concrete, much more thought is required to obtain reliable joint details.

A code-of-practice mentality can inhibit a holistic approach to design. In codes, the whole is broken down into parts that are analysed separately. The result is a safe structure but not one in which the strains and stresses have much semblance to the calculated ones. Buildings are designed for the loads expected to be applied, but rarely for the strains caused by shrinkage, creep, and temperature effects on the concrete.

Design-build contracts have meant that more consideration of the construction methods is given at the time of design. This has often led to more efficient construction (faster and/or cheaper). Unfortunately it has also meant that the design process is dominated by the demands of speed and cost. Sometimes this has led to a designer not to consider all the important effects on the final structure. The design of inadequate movement joints in buildings (e.g., car parks) is an example that has sometimes been compounded by poor workmanship.

1.1 EDGE BEAM AND COLUMN CONNECTION

A collapse occurred involving the connection of a heavy concrete gutter (1 m wide) to the supporting columns. The structure consisted of a series of reinforced concrete edge columns supporting steel trusses that spanned the width of the building. Edge beams spanned between the columns to support the concrete gutter. Figure 1.1 shows the collapse of the gutter. Although the edge beam had been designed and detailed to resist the full load from the gutter (bending moment, shear, and torsion), this had not been carried through to the joint. Figure 1.2 shows where the edge beam just started to tear away from the column and Figure 1.3 shows the top of a column where the edge beam has separated and fallen off.

Analysis — Figure 1.4 shows a failure model based on the tension strength of the concrete. Even if the links that had been detailed to pass through the joint had been constructed this way, they would have been inadequate to support the loading.

Figure 1.1 Collapse of concrete gutter.

Figure 1.2 **Edge beam starting to separate from column.**

Figure 1.3 **Column where edge beam has
broken off.**

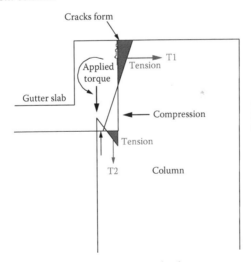

Figure 1.4 **Failure model relying on tension strength of concrete.**

A reasonable hand calculation to check the resistance is to assume that the tensile strength of concrete is about a tenth of the cube strength. In this case, the concrete strength was 20 MPa so the tensile strength may be taken as 2 MPa (no safety factors applied). Figure 1.4 shows how the tensile strength of the concrete provides the torsional resistance.

Spacing of columns	7.0 m
Supporting beam height, h_b	0.6 m
Supporting beam width, b_b	0.3 m
Width of column head	0.6 m
Width of beam support	0.1 m
Torque from self weight of gutter	$25 \times 7 \times (0.2 \times 1 \times (0.9 - 0.3/2) + 0.2$ $\times (0.8 - 0.3) \times (0.8 - 0.3)/2 + 0.3/2)$ $= 56.88$ kNm
Density of sand and bricks	18 kN/m³
Volume of sand	1 m³
Load from sand	18 kN
Volume of bricks	0.6 m³
Load from bricks	10.8 kN
Torque from sand and bricks	$(18 + 10.8) \times (0.8 - 0.3/2) = 18.72$ kNm

Total applied torque (unfactored)   **= 56.88 + 18.72 = 75.6 kNm**

$T_1 = 0.5 \times 2 \times 0.6/2 \times 0.6 \times 1000$ = 180 kN
Torque resistance of $T_1 = 180 \times 0.6 \times 2/3$ = 72 kNm
$T_2 = 0.5 \times 2 \times 0.3/(2 \times 3) \times 0.6 \times 1000$ = 30 kN
Torque resistance of $T_2 = 30 \times 0.1 \times 2/3$ = 2 kNm

Total resistance from tensile concrete   = 72 + 2 = 74 kNm

Applied load exceeds resistance

Detailing — The reinforcement for the edge beam–column joint was built as shown in Figure 1.5. The top two links were detailed to enclose

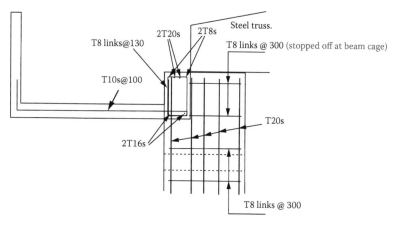

Figure 1.5 Reinforcement detailing for edge beam and column joint.

all the column main bars. Even if the links that had been detailed to pass through the joint had been constructed this way, they would have been inadequate to support the loading. If the detail had been drawn to a large scale it would have become apparent that the column links would have been difficult to fit through the beam cage, and this should have triggered concern about the connection.

Construction — The fabricator reduced the dimension of the top two links (see Figure 1.5) in order to ease the difficulties of construction.

The contractor had been using the gutter to pile bricks and sand. This load exceeded the design load and failure occurred.

Comment — Although the edge beam had been designed and detailed to resist the full load from the gutter (moment, shear, and torsion), this had not been carried through to the joint. The joint relied on the tension strength of the concrete which was not sufficient. This was a serious design error, but it is unlikely that the failure would have occurred if the gutter had been subject to just the load from rain. If the detailing and construction had been carried out thoroughly and correctly it is unlikely that the collapse would have occurred.

The collapse resulted from combined errors in design, detailing, and construction.

1.2 CONCRETE TRUSS

A project team requested a second opinion about the design of the bottom boom of a 19 m span reinforced concrete truss. The project team considered that the tension would cause unsightly cracking. A drawing (see Figure 1.6 showing the detail of part of the truss) and the analysis of the truss were provided. Much attention had been given to this design and special care had been taken to ensure that the stresses in the concrete were low.

The request was to check the design of the bottom boom. This check was carried out and it became apparent that the shear strength of the end vertical post connecting the top and bottom booms (see Figure 1.7) was inadequate. The section dimensions were $b = 230$ mm and $d = 460$ mm.

f_y	= 460 MPa
Applied shear force, V_E	= 488 kN
Design to *BS8110*[1] (*Cl 3.4.5.2*), f_{cu}	= 25 MPa

Maximum concrete strut shear capacity,
$V_{max} = 230 \times 460 \times 0.8 \times \sqrt{25}/1000 = 423$ kN (not OK)

Design to *BS EN 1992 (EC2)*[2] (*Cl 6.2.3*):
f_{cd} = 13.33 MPa
α_{cw} = 1

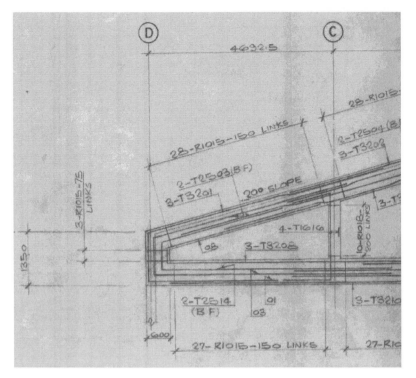

Figure 1.6 Detail of part of truss.

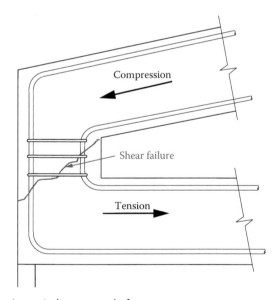

Figure 1.7 Shear in vertical post at end of truss.

z = 0.9d
v_1 = v = 0.552
cot θ = 1

Maximum shear capacity,
$V_{Rd,max}$ = 1 × 230 × 0.9 × 460 × 0.552 × 13.33/(2 × 1000) = 350 kN (not OK)

It was also apparent that the detailing required some changes. Figure 1.6 shows the main T32 bars, drawn as if they could be bent with sharp corners when in fact the minimum inner radius of bend required (3.5 × diameter of bar) is 112 mm. When drawn to scale, it became clear that the concrete would fail in two places as shown in Figure 1.8.

Failure of concrete at support (1 in Figure 1.8) — The support width was only 150 mm and the drawing did not show any reinforcement in the connection to the truss. If reinforcement was intended in the joint, then it would be difficult to fit this between the reinforcement of the truss. If not, the effect of the cover and the radius of bend of the T32 bars at the corner of the truss would lead to spalling of the concrete and the support would fail.

Failure of concrete at inner corner (2 in Figure 1.8) — Since the bottom boom was a tension member, the top reinforcement in it would be in tension. The way that it was detailed meant that a crack in the concrete would develop on the inside of the bend where it joined the vertical post at the end of the truss.

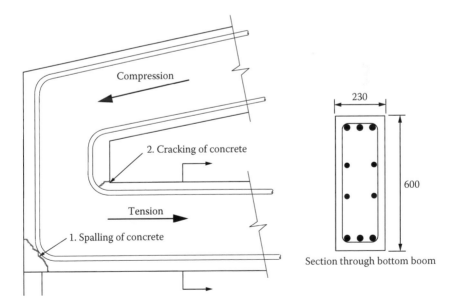

Figure 1.8 Concrete failure points.

Comment — These problems were noticed whilst checking something quite different. For the design strength of concrete, the shear reinforcement was insufficient and the concrete strut strength inadequate. If the cube strength had been increased from 25 to 35 MPa, the section would have been strong enough.

The shape of the truss was inherently unsuitable; the centre lines of the rafter and the tie did not intersect over the support. It may have led to difficult detailing if they had done so, but there would have been a good solid 'lump' of concrete to take the shear.

If the detail had been drawn to a large scale the detailing problems would have become obvious and simple changes to the design could have been made to avoid overstressing of the concrete.

1.3 CIRCULAR RAMPS TO CAR PARK

The construction of this car park (see Figure 1.9) was nearly complete. The parapet and slab of the circular ramps were supported on two columns diametrically opposite each other. The torsion and cantilever effects of the loads caused wide cracks in the supporting columns. Although these columns were damaged with shear cracks, it was considered that if the cantilever and torsional effects from the ramp could be reduced, then the columns could be repaired without risk that the cracks would return. This was achieved by inserting steel circular columns half way around at every level throughout the height of the structure (see Figure 1.10). In order to ensure that these new columns would take the correct load, flat jacks were introduced at each floor level (see Figure 1.11).

Figure 1.9 Car park with circular ramps.

Figure 1.10 Additional steel columns.

Figure 1.11 Flat jacks fitted under each new column.

All the flat jacks were connected up to a single pump and pressure gauge. Epoxy resin was pumped into each of the flat jacks at the correct pressure and allowed to harden. This ensured that the correct load had transferred to the new steel columns at each level.

Comment — There had been a design error from lack of understanding of the structural behaviour effects of the shear forces on the columns. Although the existing columns had cracked as a result of a shear failure, in this case it was not considered to be ultimate limit state and the existing columns were still capable of supporting a reduced load without risk of further damage. As soon as the error was recognised by the designer, a simple solution was provided and executed quickly without causing a delay to the overall programme.

1.4 TRANSFER BEAM WITH ECCENTRIC LOADING

A designer was considering a large transfer beam for picking up an eccentric load. The torsion design was leading to a very complicated reinforcement detail. It was discovered by a chance comment that the support for the transfer beam was on masonry and that no steps had been taken to transfer the torsion to the support. Fortunately, this error was discovered at an early stage and the whole structural design was revised.

Comment — This is a simple example in which a designer had not checked how the flow of forces was taken through the whole structure. It is unfortunate that this is not an uncommon failing.

1.5 EARLY THERMAL EFFECTS

The design of a five-storey car park incorporated long post-tensioned beams. These formed part of an unbraced concrete frame supported by 600 m square columns; floor-to-floor height was 2.7 m. The beams included six 15.6 m spans that were cast in one pour. They were stressed at 3 days to the full required prestress. The beams were 575 mm deep and 2000 mm wide, with rebates in the top corners to take 150 mm deep hollow core units. The hollow core units spanned between the beams which were at 7.2 m centres. Figure 1.12 shows the arrangement.

The construction period for the car park was extremely short (8 months) and hence all the prestressing work was carried out under a very tight schedule. Figure 1.13 shows a typical beam under construction.

Unfortunately, 5 to 6 days after the transfer of prestress for the first set of beams, cracks appeared in many parts of the frame. The cracking was

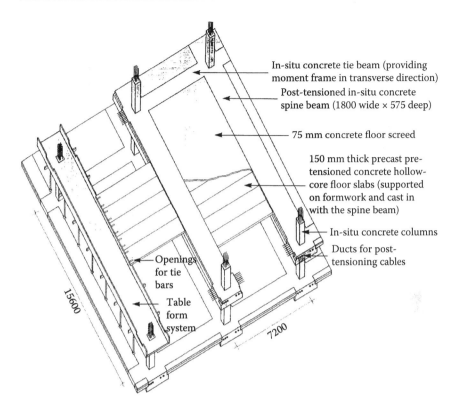

In-situ concrete tie beam (providing moment frame in transverse direction)

Post-tensioned in-situ concrete spine beam (1800 wide × 575 deep)

75 mm concrete floor screed

150 mm thick precast pre-tensioned concrete hollow-core floor slabs (supported on formwork and cast in with the spine beam)

In-situ concrete columns

Ducts for post-tensioning cables

Openings for tie bars

Table form system

15600

7200

Figure 1.12 Layout of structural members.

Figure 1.13 Typical beam under construction.

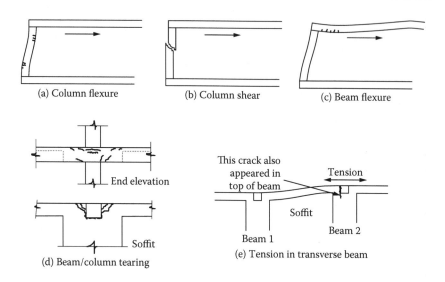

Figure 1.14 Different types of cracking.

widespread in both the columns and beams and exceeded 0.7 mm in places. Figure 1.14 shows the different types of cracking that occurred.

It took a long time to understand what caused the cracking. Whilst the problem was debated, in order to proceed with construction with minimum delay, the prestressing was altered to a two-stage process—only 50% applied at transfer and the remaining prestress applied after 2 weeks. After much discussion, it was concluded that when early thermal effects were included with the other shortening effects, the total shortening was sufficient to cause the cracking. At that time (early 1990s) it was not common to include early thermal effects in the designs of concrete frames (for reinforced or prestressed concrete). The calculated value of early thermal movement at the outer ends of a six-span beam was over 8 mm which, when added to the elastic shortening from prestress of 7 mm, provided sufficient movement to cause the cracking that occurred.

At that time, the code of practice stated 'unless the lesser section dimension is greater than 600 mm and the cement content is greater than 400 kg/m^3 there is no need to consider the early thermal effect.' Although CIRIA[3] *Report 91* covering early thermal crack control in concrete, provided a means of calculating the effect for situations of various restraints, it did not indicate what value of the restraint factor should be taken for such a beam in a structural frame. Furthermore it introduced a modification ('fudge') factor, K and suggested that this be 0.5.

The restraint to shortening of the beams by the columns was not great for this project. The early thermal movement, including frame action, was 8 mm that compared with the free movement of 10 mm. However, it would have been incorrect to base the movement on the full temperature fall from peak temperature since the beams tried to expand while heating up. A typical curve for one of the beams, giving the temperature rise and fall over time, is shown in Figure 1.15.

From a layman's view, as the concrete started to set, the temperature rose. During this period, the concrete remained somewhat plastic. The stiff columns prevented any expansion of the beams (they just bulged a bit). When the temperature started to fall, the concrete had hardened and the beam shortened, causing the cracks.

Comment — This is an example where the effects of early thermal movement should have been checked because of the long continuous multi-spans. The beams happened to be prestressed but the same effects apply to reinforced concrete structures. Typically, for a 300 mm internal floor, an allowance of 100×10^{-6} for early thermal contraction strain should be made. This project provided the first clear record of early thermal effects in a concrete frame structure. The results emphasized the importance of including these effects in design—not common practice before 1995.

This particular contract was complicated by the fact that it was a design–build type. The design of the beams was carried out by a specialist subcontractor, but the responsibility for the frame design was with the structural engineering consultant. This split responsibility contributed to the confusion and delay in resolving the problems. In such contracts *it is essential to name a single designer or engineer who retains overall responsibility for the stability of the structure, the compatibility of the design, and details of the parts and components, even where some or all of the design including detailing of those parts and components are not carried out by this engineer.*

Reference should be made to *Design of hybrid concrete buildings.*[4]

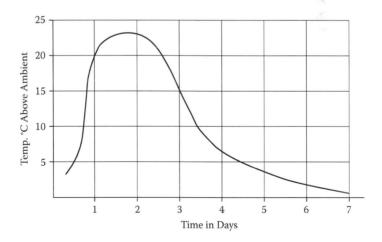

Figure 1.15 Typical early temperature rise and fall in concrete beam.

1.6 SECONDARY EFFECTS OF PRESTRESSING

A concrete frame with a two-span prestressed beam had been designed for serviceability limit state (SLS). When checking the frame for the ultimate limit state (ULS), the designer discovered that the moments induced into the edge columns exceeded their capacity, 850 kNm. Figure 1.16 shows the layout of the structural model of the frame. The moments from the ULS analysis are shown in Figure 1.17. To reduce the transfer moment at the edge, the designer considered creating hinges between the column and slab as shown in Figure 1.18. This proposal was not favoured because:

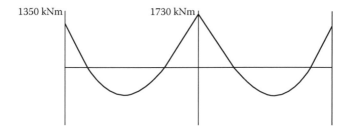

Figure 1.16 Layout of structural model of concrete frame.

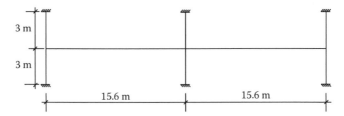

Figure 1.17 Moments from ULS analysis.

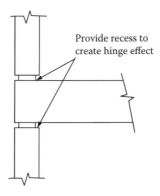

Figure 1.18 Proposal for providing hinges.

- Reinforcement must be placed centrally to allow a hinge effect to be created.
- Compression in the concrete section would automatically create moment transfer.

The question raised was whether secondary effects had been included in the ULS analysis. To answer this question, it is important to understand the method of analysis. For prestressed structures in the UK, it is common to carry out the primary analysis at SLS. Often the equivalent load method is used. For this the prestressing effects are modelled as an equivalent load (e.g., a uniformly distributed load models the prestressing effect of a parabolic drape of the tendons). This type of analysis takes account of secondary effects. For those not familiar with this type of work, secondary (parasitic) effects can be summarised by the diagrams shown in Figure 1.19.

A check is then carried out at ULS. For this limit state, the prestressing action is often considered as affecting only the resistance and not as part of the loading. The secondary forces and moments must be calculated separately and added to those of the primary analysis. For this particular frame, the secondary bending moments are shown in Figure 1.20. These effects had not been included in the ULS analysis check for this project. When added to the existing results, the moments became acceptable as shown in Figure 1.21. Once it was agreed that this was the correct analysis, it was realised that the column was not overstressed and the moments in the span of the beams had increased. There was plenty of room in the section to add more reinforcement to take account of this.

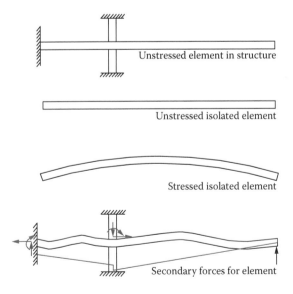

Unstressed element in structure

Unstressed isolated element

Stressed isolated element

Secondary forces for element

Figure 1.19 Example of prestress secondary effects.

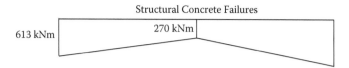

Figure 1.20 **Secondary effects for frame.**

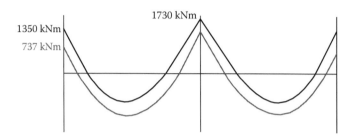

Figure 1.21 **Moments from ULS analysis adjusted for secondary effects.**

Comment — This problem was partly due to the design method used in the UK. For prestressed beams, the primary design/analysis is often carried out at the SLS. The prestress is considered a load (or action effect) within the equivalent load method.

It is not possible to use this approach for the ULS check as the prestressing tendons resist the applied actions in the critical parts of the beam and thus do not provide an equivalent load. The secondary effects must be considered separately at such parts. The debate about the interaction of secondary effects and moment redistribution is ongoing. However, in this situation there was no reason to believe that moment redistribution would or could take place before damage occurred to the column.

1.7 TEMPERATURE EFFECTS ON LONG-SPAN HYBRID STRUCTURE

The ground level of a two-storey underground car park slab was not covered. The structure consisted of 16 m spanning hollow core units bearing on precast concrete beam nibs. Movement joints had been shown on the drawings but these did not function correctly for a variety of reasons. The upper surface of the slab was exposed to the weather and in particular to large variations in temperature. The latter caused movement and rotation of the units and their supports. This resulted in severe cracking of the supporting nibs and in some places cracking at the ends of the hollow core units. Even after repair, the cracks reappeared each subsequent year for

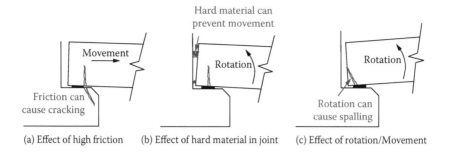

(a) Effect of high friction   (b) Effect of hard material in joint   (c) Effect of rotation/Movement

Figure 1.22 **Examples of potential failures at movement joints.**

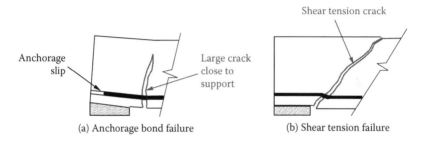

(a) Anchorage bond failure   (b) Shear tension failure

Figure 1.23 **Types of end failure of hollow core units.**

more than 5 years. Figure 1.22 describes different mechanisms that can occur even where a structural topping has been used.

In general, if the bearing material creates large friction forces, these can lead to large tension stresses in both the support and the precast slab or beam (see Figure 1.22a). Neoprene bearings or similar should be used to avoid this.

If the space between the precast slab or beam and the face of the supporting member is not adequate for the required movement or it fills with hard material over time, cracking will occur (see Figure 1.22 b). If the effects of movement and/or rotation cause the line of action to move too close to the edge of the support, local spalling can occur (see Figure 1.22c).

A further problem arises where hollow core units are used. Any cracking that occurs close to their ends is likely to cause anchorage bond or shear failure of the unit (see Figure 1.23). Anchorage bond failure may occur because the cracking close to the support does not allow the full anchorage resistance to develop and the prestressing strands start to slip. This causes the crack to grow until the unit fails (see Figure 1.23). Hollow core units are inherently vulnerable to the effects of cracking close to the support as their shear resistance relies on the tension strength of the concrete. Unlike a solid section, a cross section available for shear resistance of these elements is much reduced due to the presence of the cores.

Comment — The remedial work to correct this sort of problem is likely to be very expensive. Making good the cracks is not a satisfactory solution as the cracks reappear on an annual cycle. Eventually the danger of falling concrete to users of the car park results in temporary works that may be so extensive that a rebuild of the whole structure can become a sensible solution.

Where precast concrete elements form a major part of a structure, careful calculation of tolerances is essential. This was particularly so for this project as the design of the movement joints did not take tolerances into account sufficiently. This aspect of design is not so critical for in situ concrete structures as many of the tolerances get absorbed satisfactorily by the construction process.

This project was another example showing *it is essential to designate a single designer or engineer who retains overall responsibility for the stability of the structure and the compatibility of the design and details of the parts and components, even where some or all of the design including detailing of those parts and components may not be carried out by this engineer.* Reference should be made to *Design of hybrid concrete buildings*[4] and the Concrete Society's report titled *Movement, restraint, and cracking in concrete structures.*[5]

1.8 LOADING FOR FLAT SLAB ANALYSIS

A flat slab spans areas between columns and failure can occur by the formation of hinges along the lines of maximum hogging and sagging moments. This can be most easily presented using the folded plate theory as shown in Figure 1.24. A complementary set of yield lines can form in the orthogonal direction.

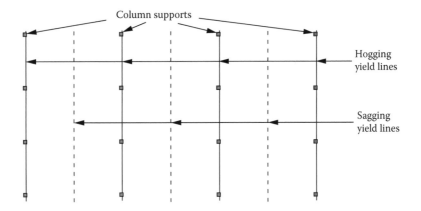

Column supports

Hogging
yield lines

Sagging
yield lines

Figure 1.24 Simple yield line mechanism for flat slab.

One misconception of some engineers is to consider a reduced loading when analysing a flat slab. Each moment applied in each orthogonal direction must sustain the total loading to maintain overall equilibrium. There is no sharing of the load by partial resistance in each direction.

Comment — One engineer, who had worked in the U.S., found that this erroneous belief went back to the early days of flat slab construction, when the promoters measured the strains in the reinforcement of newly constructed slabs. They ignored the contribution of the concrete in tension and promulgated the idea. The engineer had, in fact, designed some warehouse slabs by that method. Many years later, a new owner asked whether the imposed floor loading could be increased, and one of his colleagues checked it and found that it was overloaded under the dead load.

Although the author cannot quote other instances where this occurred in designs, there have been many comments by engineers to this effect. It is clear that many engineers believe that it is reasonable to consider a reduced loading. Reference should be made to the Concrete Society's technical report titled *Guide to the design of reinforced concrete flat slabs.*[6]

1.9 PRECAST CONCRETE CAR PARK

The layout of the car park deck included precast spine beams supporting double tee units. Figure 1.25 shows a section through the deck close to the edge of the structure. A structural screed was placed over the whole deck. The method of construction for the cantilever units was unconventional and included on-site welding. Figure 1.26 shows an enlarged detail through the section.

The structural screed laid over the whole deck was designed with a step over the welded plate above the spine beam. The final screed, incorporating a waterproof membrane, was laid to falls. The drainage points were situated at the ends of the spine beams. Figure 1.27 shows the final arrangement through a section. The step in the screed created a crack inducer for the tension stresses at the top of the cantilever and water draining off the

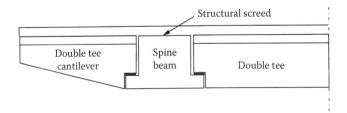

Figure 1.25 **Section edge of car park deck.**

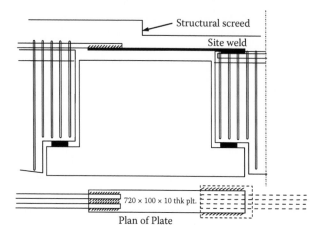

Figure 1.26 **Detail through section.**

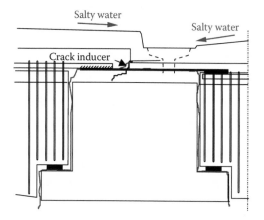

Figure 1.27 **Section through completed deck.**

car park was likely to permeate it. In winter, this water was likely to be combined with de-icing salts. The risk of corrosion of an important part of the structure was high. By the time this problem was realized, it was considered too late to change the design and form of construction. Instead the depth of the waterproofing membrane was increased.

Comment — The design included four major errors:

1. Site welding for the main cantilever reinforcement is not recommended. When checked on site, some of the welds were found to be of very poor standard and had to be condemned.

2. The design included filling the spaces between the spine beam and double tee units with mortar. This was very difficult to achieve and resulted in poor compaction that allowed water passage. The design should have provided a better solution.
3. The sharp corner in the structural screed formed a crack inducer.
4. There was a high probability that salt water would penetrate the structure and cause corrosion of the reinforcement.

If the membrane was asphalt or similar (made to carry wheel loads) and fully bonded to the screed, it would with time become so brittle that the crack in the screed would propagate into the membrane.

1.10 ARCH FLOOR

The construction of a hotel was nearing completion. The first level floor spanning 10 m had been designed as an in situ concrete flat arch. The design kept the floor thickness at mid span to a minimum (100 mm) increasing toward the ends to 275 mm thick at the support. The formwork for parts of the floor had not been supported sufficiently and deflected to cause the floor thickness to be 75 mm thicker than designed. The slab was supported on masonry walls. Some months after construction, cracks were noticed in the floor and were considered to have been caused by shrinkage. The client, for other reasons, complained about the construction and an independent engineer was asked to give a second opinion. During this inspection, it was discovered that no design ties to the flat arches had been provided, although no significant movement had occurred at the supports.

A number of remedies were considered and the one that appeared to be most suitable was to insert Macalloy bars and anchor them to the outsides of the supporting masonry walls. Special fire protection from boxes made of fire-resistant material was required.

Comment — It was by pure chance that the structural integrity of the building was called into question and it is quite possible that it would not have failed or collapsed for many years, if ever!

1.11 PRECAST CONCRETE STAIRFLIGHTS

The 1968 collapse of a staircase during construction, killing two people, caused a rethink in design, detailing, and construction. The precast stair-flights were designed with half joints to sit on in situ concrete landings. One of the upper floor stairflights had been placed in position by a crane but it needed to be shifted into its correct position. Temporary 'acro' props were used from the flight below to support and lift it as it was levered into its final position. The stairflight below was supported on an in situ concrete nib that had been cast only a few days earlier. This nib did not have its full design strength and was loaded with almost twice its design load when failure occurred. The thickness of the landing slab was only 200 mm which meant that the depth of the nibs, 100 mm, did not allow sufficient room for reinforcement.

Since that time, precast manufacturers have developed a variety of different types of joints using precast stairflights. When using such a proprietary system, it is essential that a designer consider:

- The method of adequately tying the stairflight to adjacent parts of the structure
- The sequence of construction
- The temporary works involved
- The chain of responsibility in achieving the final structure (Often the temporary loads due to props and other factors can provide the critical design condition.)

Many current systems do not include adequate ties between the stairflight and the adjacent structure. The following examples show how tying reinforcement can be provided.

Half-joints — Figure 1.28 shows a typical layout of reinforcement where half joints are used. The tying reinforcement that projects from the precast units provides continuity with the reinforcement in the structural screed.

Dowel joints — To provide sufficient room for a dowel hole, the dimensions of the nib must be increased to those shown in Figure 1.29. Figure 1.30 shows a preferred arrangement of reinforcement for a dowel connection.

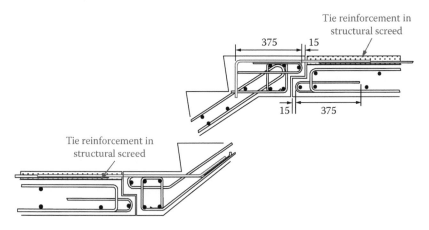

Figure 1.28 **Typical layout of reinforcement for half joints.**

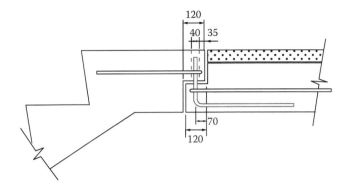

Figure 1.29 Half-joint with dowel.

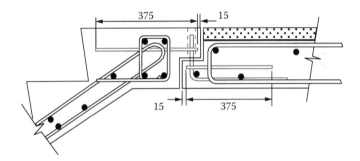

Figure 1.30 Reinforcement arrangement for dowel connection.

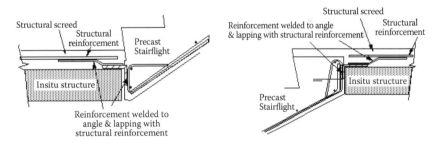

Figure 1.31 Stairflight system using steel angles.

Proprietary system using steel angles — Figure 1.31 shows how a proprietary system can be adapted to allow continuity of ties.

Comment — This failure has been included within this chapter although it is clear that the detailing and construction faults were important contributors. Great care is required when constructing staircases with precast stairflights. The importance of this cannot be overemphasised. A hybrid form of construction is commonly used and the dangers get easily forgotten. Although safe construction largely depends on workmanship, a designer has an important obligation to provide a robust and buildable solution. Reference should be made to *Design of hybrid concrete buildings*.[4]

1.12 SHEAR STUDS ON STEEL COLUMN TO SUPPORT CONCRETE SLAB

The designer's proposal was to use shear studs welded to a steel column to transfer the load from the concrete flat slab to the column (see Figure 1.32). However, with this configuration, a punching shear failure could occur with only the bottom row of studs providing any shear resistance. All the other shear studs would remain in the cone of concrete attached to the column (see Figure 1.33). To avoid such a failure, two options provide sensible solutions:

Option 1: Provide links to transfer the load back to top of slab — Links, in addition to those required to resist punching, would be required to transfer the full load to the top of the slab. Struts in the concrete would then transfer the load on to the shear studs (see Figure 1.34). This is unlikely to provide the most practical option.

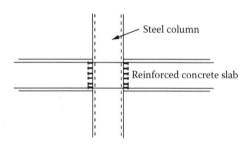

Figure 1.32 Junction between steel column and concrete slab.

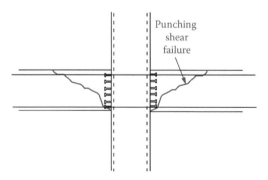

Figure 1.33 Punching failure of slab.

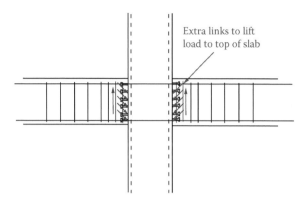

Figure 1.34 **Use** of links to transfer load to shear studs.

Option 2: Provide shear key (instead of stubs) at bottom of slab — Using a shear key with full strength welds at the bottom of the slab would provide adequate resistance. The shear links would ensure against a punching failure and the resulting force flow would engage the shear key (see Figure 1.35). The size of the shear key would depend on the thickness of slab and the shear force. The upper limit to this solution would be the compression strength of the concrete at the column face:

BS 8110[1]: Maximum compressive stress = $0.8 \sqrt{f_{cu}}$ but not more than 5 MPa
BS EN 1992[2]: Maximum compressive stress = $0.8 \times 0.6 \ (1 - f_{ck}/250)$ MPa

Comment — This is an example where thinking struts and ties can be helpful.

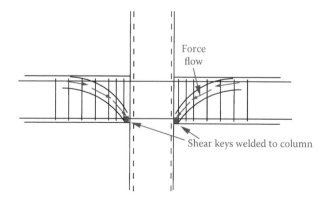

Figure 1.35 **Use** of strut-and-tie model to transfer load to shear key.

1.13 PILED RAFT FOR TOWER BLOCK

The design of the raft assumed that the walls of the two-level basement car park would act with the raft over the piles to transmit the shear and bending forces to the outer piles. The walls of the basement had almost full height openings, placed one above the other, and contained only nominal reinforcement. The combined strength of these walls plus the 1.5 m thick slab was inadequate to transmit the loads (see Figure 1.36).

The mistake was discovered whilst the tower block was being constructed. The remedial work required a new raft to be constructed beneath the existing one (see Figure 1.37). The new design relied on the composite action of the new and old rafts. All the existing surfaces of concrete were scabbled and further reinforcement was laid between the existing piles. Placing of concrete for the lower part of the new raft was carried out conventionally.

To achieve good bond with the bottom of the existing raft, the upper part of the new raft was packed with single sized aggregate and then grouted with a retarded and fluid cement paste. The grout was introduced under pressure to the back of the pour through a complicated system of metal pipes pinned to the underside of the existing raft. The method produced a wall of grout that extended from top to bottom of the pour and flowed forward toward the peripheral shutters with the top surface behind the bottom. Pipes were so placed to let the air out in front of the grout surface, then indicate where and when the grout arrived, and then allow grouting to continue from immediately behind the advancing wall of grout. Grouting was continuous until the work was complete.

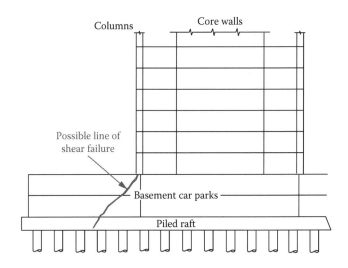

Figure 1.36 Arrangement of piled raft and basement car parks.

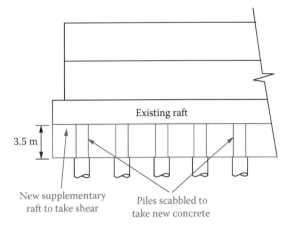

Figure 1.37 **Schematic arrangement of new raft.**

Comment — This error was discovered as a follow-up to a check of some of the reinforcement drawings of the tower block that were found to contain mistakes.

1.14 FLOATING PONTOON FOR RESIDENTIAL BUILDING

The design and construction of a floating pontoon to carry a residential block were complete. The pontoon was afloat in the marina, ready for work to start on the building. As the load increased, it became apparent that the pontoon was not sufficiently buoyant. The design check that followed showed that the density of concrete assumed was too low. This error was compounded by errors in construction (oversizing of the walls and base of the pontoon), all of which resulted in a serious reduction in buoyancy. The situation was resolved by attaching large blocks of polystyrene to the concrete structure of the pontoon.

Comment — During the design and construction of the pontoon, insufficient thought had been given to the essential requirement of ensuring that the structure was adequately buoyant.

1.15 PRECAST COLUMN JOINT DETAIL

A precast sloping column, 400 mm in diameter, was designed to take the load from several floors above (see Figure 1.38). A short steel stub was cast

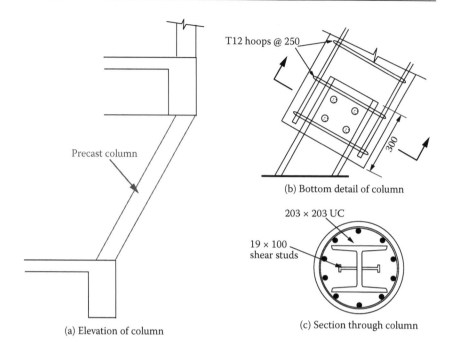

(b) Bottom detail of column

(c) Section through column

(a) Elevation of column

Figure 1.38 **Sloping precast column.**

into the bottom of the precast element to simplify the connection detail to the beam below. During construction, it was realised that the design of this connection was unsafe due to insufficient bond capacity and anti-bursting steel.

The short stub consisted of a 254 × 254 UC section that extended into the column about 300 mm. The stub had four 19 × 100 shear studs welded to each face of the web (see Figure 1.38b). The links surrounding the stub column were T12 hoops at 250 mm pitch. No other anti-bursting reinforcement had been provided.

The columns were already constructed and supported four floors. Check calculations showed that the factor of safety against collapse was about 1.1 and this would reduce to less than 1 when the next floor was added 4 days later. Work was stopped in that part of the project and temporary props installed in the area to prevent possible collapse.

The remedial action was to encase the existing column with a steel circular hollow section, 457 mm diameter, of the same length. This was cut into two halves along its length and then welded in place with full strength welds along the full length. The space between the steel and precast columns was then grouted up (see Figure 1.39).

No calculations had been carried out to determine the required embedment length of the UC stub column. Bond and end bearing should have been considered.

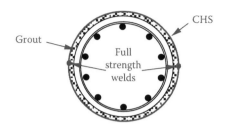

Figure 1.39 **Remedial work.**

Bond — If the force was to be transferred using bond stress, a design shear strength of 0.6 MPa independent of concrete strength should have been used. This is the value given in BS EN 1994-1-1[7] for the design shear strength due to bond and friction for a completely concrete-encased section. The design value of studs and shear connectors may also be taken from that standard. The concrete around shear connectors fixed to the web of H sections is partially constrained by the flanges, and this can result in higher design resistances.

End bearing — Some of the column force is taken in end bearing. The bearing stress should be limited to three times the design concrete compressive strength, e.g., $3 \times 0.85 \, f_{ck}/1.5 = 1.7 \, f_{ck}$ for designs to BS EN 1992-1-1.[2]

Links — Links must be provided over the embedded length to prevent splitting. The force to be resisted by the links (one leg) in a unit length should be equal to the shear force transferred in that length divided by 2π. The links above the stud must resist a force equal to the end bearing force divided by 2π.

Comment — It was a fortunate chance that this fault was found. The engineer who discovered it was intending to copy the detail for a similar nearby structure.

Chapter 2

Problems and Failures due to Errors in Structural Modelling

2.1 REINFORCED CONCRETE TRANSFER TRUSS

A multi-storey building constructed according to a design–build contract included a deep transfer reinforced concrete truss. The client required a second opinion on the design of this from an independent consulting engineer. The report of this engineer condemned the design as unsafe. The contractor then employed another consulting engineer to carry out a further check and, if it confirmed the findings of the report, to find a solution that provided the least disruption to the existing works.

The second check revealed that the initial design of the concrete truss involved the use of a complicated finite element program. The end support to the truss was a stiff wall and this had been modelled with plate elements that had no lateral stiffness (see Figure 2.1). The second consulting engineer then carried out separate frame analyses with simpler programs that included more realistic properties for the edge wall. The results showed that the design of the trusses was safe—just. The reinforcement details of the nodes required some minor changes.

Comment — This example shows that the original designer had not understood the effect of the end wall stiffness or did not understand the limitations of the modelling software. Both cases reflected some incompetence. This is an example in which the two checking engineers acted for parties with opposite views about the safety of the structure. It shows how vulnerable engineers can be in trying to prove what they are asked to do rather than examining the information objectively.

2.2 MODELLING RIGID LINKS

It is possible in some structural frame programs to set up rigid links in the data. These allow the user to specify that part of the structure that will behave rigidly between stated nodes. A rigid link may include any number

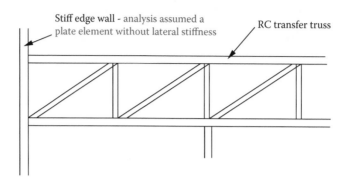

Figure 2.1 Diagrammatic view of concrete truss.

of elements from a single pair to a large group representing a stiff part of the structure; for example, all nodes on a particular plane may be linked to represent the in-plane stiffness of a floor slab without explicitly modelling it. It is often possible to link only some of the degrees of freedom; for example, with the floor slab, the out-of-plane translation and the in-plane rotations must not be linked.

For the design of one particular building, rigid links were used to model floor plates. At a level where the wall arrangement within individual cores changed significantly, the rigid links had to translate very large forces through the floor slabs (transferring forces from one core structure to another). The program did not provide any output of the forces within the rigid links and the designer assumed that loads could be transferred through the floors safely. During a design review, it was pointed out that these forces could be very large. In fact, it was then discovered that the design forces exceeded the design resistance of the materials and a redesign had to be carried out during construction of the building.

Comment — This example highlights the need to know what level of information is required and what level of detail will be provided by a structural model. In this case, the designer had not given thought to the problem and the structural modelling system did not supply adequate information.

2.3 ASSESSING MODEL LIMITS AND LIMITATIONS

Most analytical models for reinforced concrete, including those for ultimate limit state (ULS), are based on the elastic properties of the concrete section, uncracked and without reinforcement, i.e., a homogeneous material. This allows a simple analysis of a frame or structure with uniform stiffness for each element, and for most situations provides an acceptable

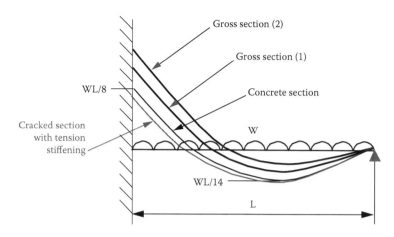

Figure 2.2 Design analysis of propped cantilever.

solution for design. The reason this is normally acceptable can be explained by considering a simple indeterminate structure—a propped cantilever slab with UDL. Figure 2.2 shows the flexural results of analysis for several different situations.

The first analysis assumes simple concrete section properties—a homogeneous section throughout. This gives a maximum support moment of WL/8 and a maximum span moment of WL/14. The engineer then designs the reinforcement for these moments and, of course, the reinforcement required at the support is nearly twice that required in the span. The effect of including this reinforcement (giving gross section properties) is to increase the stiffness of the slab near the support.

If a further analysis is then carried out with the revised stiffness properties (not normally done), the bending moment is altered to that showing Gross Section (1). This would cause the engineer to increase the reinforcement at the support and reduce it in the span, causing a further increase of stiffness at the support and reduction in the span. The next analysis would result in moments shown as Gross Section (2). This iterative process would lead the engineer to design the slab as a cantilever. However, in reality, the slab will crack under the ultimate loads. This reduces the stiffness at the cracked sections. A cracked analysis of the section with the reinforcement required by the analysis using simple concrete section properties would result in moments shown as Cracked Section with Tension Stiffening. This results in a similar curve to that shown for the analysis using simple concrete section properties

If, as a result of the first analysis with concrete section properties, the neutral axis depth at a critical section is greater than that for a balanced section (i.e., the reinforcement does not yield at the ultimate load), and the applied moment is less than the cracking moment, there is a possibility of a premature brittle failure. Both these conditions should be checked to ensure that the critical sections are capable of moment redistribution. Otherwise the design may be unsafe.

Many reinforced concrete continuous beam and slab and frame programs model the elements with homogeneous properties and the engineer inputs these with just the concrete section without reinforcement. It is thus important that the software, after it designs the reinforcement, contains checks for neutral axis depth and cracking to ensure against premature brittle failure.

This demonstrates that design is heavily reliant on a material's capacity to redistribute the forces achieved by two independent means: (1) cracking of the concrete, and (2) yielding of the reinforcement. The effect of these can often provide a very different ultimate resistance of a structure than that predicted from the elastic results.

Although elastic results are normally conservative, in some modelling situations this is not so. A typical example is in the use of a plane frame program for analysing flat slabs. The modelling simplifies the flat slab as though it is a continuous slab spanning on continuous supports (walls). The bending moments from this analysis peak over the supports but do not vary across the width of the slab. In fact, the bending moments peak over the actual column supports so the total bending moment calculated for the full width of slab should be distributed across the line of support, unevenly peaking over the support. The code provides rules to compensate for this effect (using column and middle strips). An incorrect distribution of moments will also cause incorrect distribution of shear forces.

2.4 EMPIRICAL METHODS

The development of good practices has relied on empirical and more rigorous analytical methods. Many empirical methods have been honed from experience to provide very efficient (cost effective) solutions. Sometimes this has caused these methods to be less conservative than a more rigorous approach.

One reason for this is that the material properties of the concrete and reinforcement have changed since the empirical methods were introduced and the original assumptions for them are no longer valid. This has caused some concern and discussion amongst the code writers. Generally it is felt that the empirical solutions should have an overall built-in safety factor that is greater than that for more rigorous methods. Over recent years this has been implemented for ultimate limit states by editing the code clauses.

2.5 INITIAL SIZING OF SLABS

During the past 50 years, the reduction of partial safety factors and increases in strengths of materials have resulted in the thickness of slabs often being controlled by serviceability limit state (SLS; deflection and cracking). However, determining accurate deflection information of reinforced concrete slabs and beams at the design stage is almost impossible. It requires knowledge of the following factors and/or properties:

1. Materials used (type of aggregate, type of cement, and amount of water)
2. Weather conditions at time of construction (hot or cold, humid or dry)
3. Mechanical properties of hardened concrete (compression and tension strength, modulus of elasticity)
4. Mechanical properties of reinforcement and characteristics of its bond with the concrete
5. Actual loading on the element, not only at the point in time required, but also the load history up to that moment
6. Effect of continuity with adjacent structural element
7. Exact construction process and sequence

Where accurate information exists, for example, when checking an existing structure, it is possible to make an excellent calculation assessment of the deflection (certainly within 5 mm).

In contrast, at the time of design, very little accurate information exists about concrete properties or loading. Nevertheless, for the purposes of design, codes of practice make reasonable assumptions for all the above factors and properties and provide simplified methods to aid the design engineer. The most important of these is the span/effective depth method. However it is important to realise that this method cannot be considered more than a rough guide and those who rely on it to pare the design thickness of a slab over 7 m in span to an accuracy of less than 25 mm, without the benefit of experience or specific information, are living in 'cloud cuckoo land'! It is thus wise to err on the conservative side when using the span–effective depth method to assess the design depth of a slab.

2.6 ANALYSIS OF FLAT SLABS WITH FINITE ELEMENT PROGRAMS

When using finite element programs to analyse and design flat slabs, it is important to realise that to reinforce for the peak moments over the

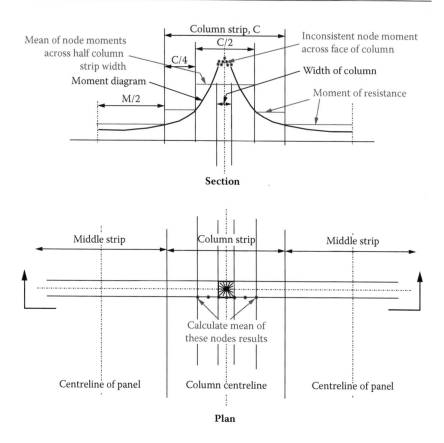

Figure 2.3 Bending moment results from typical finite element program.

supports is normally considered overly conservative. Common practice assumes some lateral redistribution of the peak moments. Figure 2.3 shows the moment profile results across the full width of a typical slab at the support from the finite analysis. Often the moments around the support node are inconsistent. It is reasonable within half the column strip (the column strip is defined as half the total width of slab) to assume mean moments across this width. This may cause cracking under working conditions and one or two of the reinforcing bars over the column may even reach their yield stress. This is normally considered acceptable but a designer should be careful to ensure that the analysis is satisfactory for the particular situation.

This is one example in which a designer relies on the ductility properties of a slab. This ductility results from both cracking of the slab and yielding of the reinforcement. The cracking of the slab causes its stiffness to reduce and hence the redistribution of moments. The yielding of the reinforcement causes a plastic hinge with similar effects.

The choice of the stiffness for a flat slab and supporting columns for ULS analysis depends on engineering judgement. Taking half the uncracked concrete section properties for the slab and the full uncracked concrete properties for the columns is considered reasonable for most situations. This is discussed in more detail in the Concrete Society's *Guide to the design of reinforced concrete flat slabs.* [6]

2.7 SCALE EFFECTS

Design engineers sometimes find that they are involved in very large size structural projects (e.g., multi-storey blocks, heavy transfer slabs, deep rafts on large piles, etc.). There is a temptation to continue to use the rules of thumb and empirical methods that are common for small and medium sized structures. Often these or the assumed simple reinforcement detailing layout are not applicable and may lead to unsafe designs.

It is important to think more fundamentally how the forces are transmitted and ensure that reinforcement is provided and fully anchored where tension forces need to be resisted. A typical example would be a pile cap that has 3 m diameter piles and requires a thickness over 5 m. The strut forces must be considered carefully and the nodes reinforced to ensure the concrete can contain the forces. This can lead to design reinforcement in addition to the normal bottom bars that enclose the cap to resist the bursting forces.

Strut-and-tie methods are very useful in helping explain the flow of forces but they are often difficult to apply quantitatively. *BS EN 1992-1-1* [2] provides some helpful rules for such applications. When dealing with deep sections, it is important to remember that the lever arm for calculating the tension in the bottom reinforcement must not exceed 0.6 times the span (maximum height of the arch).

Chapter 3

Failures due to Inappropriate Extrapolation of Code of Practice Clauses

During the 1960s and 1970s, there was a strong inclination amongst some engineers to extrapolate clauses in the codes of practice beyond reasonable limits. This may have been a result of the growing pressure to increase spans without reducing the depth of floors and without adding further costs to designs. Rule-of-thumb methods had been developed and refined for floor spans up to 7 m but increasing demand for longer spans made it all too easy for engineers to extrapolate the current methods without ensuring that they were still valid.

Codes of practice give rules that can at best provide approximations and will work in most, but not all cases; they do not necessarily give factual information.

3.1 COOLING TOWERS

In November 1965, three out of eight cooling towers collapsed in gales of over 85 mph (see Figure 3.1). Each tower was 375 feet high and they had been constructed closer together than usual and had greater shell diameters and shell surface areas than any previous towers. The high winds were considered to have triggered the collapse, but an inquiry found the exact cause to be an amalgamation of several other factors in the tower design:

- British Standard wind speeds had not been used in the design; as a result, design wind pressures at the tops of the towers were 19% lower than they should have been.
- Basic wind speed was interpreted and used as the average over a 1-minute period, whereas, in reality the structures were susceptible to much shorter gusts.
- The wind loading was based on experiments using a single isolated tower. The grouping of the towers created turbulence on the leeward towers—the ones that did actually collapse.
- Safety margins did not cover uncertainties in the wind loadings.

Figure 3.1 View of collapsed cooling towers. (Courtesy of Gillian Whittle.)

Based on the findings, wind loading in the initial design was seriously underestimated.

The code of practice used at that time was *CP 114*[8]. It used overall safety factors and did not provide factored loads with special combinations. The collapse occurred a month before the publication of a paper by Rowe, Cranston, and Best on *New concepts in the design of structural concrete.*[9] This new approach to design would have ensured that the design loading combination would have resulted in a safe structure.

Figure 3.2 shows a simplified view of the overall design forces, where W is the wind load; G is the self weight of the cooling tower; h_w is the height of the centre of the wind force; and b is the width of the base of the cooling tower. The two design approaches give very different requirements to ensure that no overall tension forces are created in such structures.

Permissible stress Code (*CP 114*) approach:
  Applied wind load moment = $W \times h_w$
  Restoring gravity moment = $G \times b/2$
$$G \times b/2 \geq W \times h_w$$

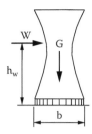

Figure 3.2 Equilibrium requirement.

New limit state approach with factored loads:
Applied wind load moment = $1.4(W \times h_w)$
Restoring gravity moment   = $1.0(G \times b/2)$
$$1.0\ (G \times b/2) \geq 1.4\ (W \times h_w)$$

Comment — This incident should really be explained in a chapter of its own as the failure was a result of a philosophy built into the then current code of practice, *CP 114*. Although there were some defects in wall thickness, the main cause of the collapse was because the design value chosen for the wind load was too small. This collapse ensured the early adoption of a limit state code of practice in the UK, resulting in a completely new approach to design. The first draft of the unified code appeared in 1968 and in 1972 it was published as *CP 110*.[10] This was the first comprehensive limit state code of practice ever published.

3.2 DESIGN BENDING MOMENTS

It is common for engineers to use simplified coefficients to obtain bending moments for continuous beams and slabs. Although it is generally conservative to design continuous beams with pin supports (i.e., no moment transfer to the supports), it may be unsafe to assume that no moment is transferred to the columns or end supports of the beams. As an example, designers using *BS 8110*,[1] Clause 3.4.2 and Table 3.5, often incorrectly assume pinned supports.

The coefficients given in Table 3.5 of the standard are unsuitable for situations where moments can be transferred to the supports. This is particularly so for end supports. Fortunately, the detailing rules require some reinforcement to resist nominal hogging moments at the ends of beams. It is not uncommon for engineers to ignore any moment transfer to supports even when the adjacent spans differ by more than the limitations given in the codes (e.g., variations in span length should not exceed 15% of the longest). Even masonry supports provide some moment resistance.

The design of a particular concrete frame included a two-span beam for which one span was twice the length of the other. The analysis for vertical loads assumed that moments were not taken into the internal column. This was questioned at a design review. A check showed that the column moment capacity was not sufficient to take the moments caused by the beam rotation at that support. It was fortunate that the design loads could be reduced sufficiently so that a redesign of the column was not required.

Comment — Although it is not possible to cite particular failures from this case, it is another example where the ductility of reinforcement and the redundant capacity of most structures allow alternative load paths to prevent failure.

3.3 PILES WITH HIGH STRENGTH REINFORCEMENT

An engineer was hoping to reduce the number of piles required for a building by providing reinforcement with much higher strength than required for normal high yield bars. Figure 3.3 shows a section through a 225 mm diameter pile with a proposed Dywidag threaded bar 36 mm in diameter (Dywidag Systems International, Munich). The yield stress f_y of such a bar is 1200 MPa.

Before going into production it was decided to check the load carrying capacity with a test pile. The result was that the pile failed at a similar load to that of a pile with normal reinforcement with f_{yk} of 460 MPa.

After some thought, the reason for the unexpected low pile strength became apparent. The actual concrete compression stress–strain diagram is similar to that given in Part 2 of *BS 8110*[1] and Figure 3.2 of *BS EN 1992 (EC2)*[2] and differs in shape from that used in ULS design in that, instead of having a plateau of maximum stress, it peaks at about 0.002 strain. This strain is close in value to that of the yield strain of normal reinforcement $(500/200000 = 0.0025)$. This means that the maximum resistance of the concrete occurs at the same strain when the reinforcement reaches its yield stress (maximum resistance).

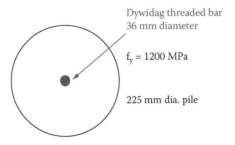

Dywidag threaded bar
36 mm diameter

f_y = 1200 MPa

225 mm dia. pile

Figure 3.3 **Proposed pile with Dywidag bar as reinforcement.**

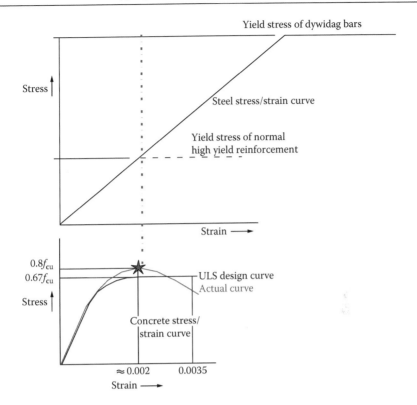

Figure 3.4 Stress–strain curves for concrete and steel.

The yield stress of a Dywidag bar is about 1200 MPa which corresponds to a yield strain of 1200/200000 = 0.006. This compares with the ultimate strain of the concrete of 0.0035, so the concrete will fail in a compression test long before the Dywidag bar reaches its yield strain or stress. Figure 3.4 shows this graphically.

Comment — This is an example where extrapolating from the code clauses (*BS 8110*, Table 3.1) can lead to unexpected results. It is important to understand the limits assumed in the code, if they are likely to be exceeded in a particular design.

3.4 SHEAR CAPACITY OF DEEP SECTIONS

The design of a prestressed bridge section followed a code of practice that some considered to overestimate the shear capacities of prestressed beams. Soon after construction, cracks appeared. There was a long debate about the cause of cracking and, until remedial work was carried out, temporary additional external shear reinforcement was put in place (see Figure 3.5).

Figure 3.5 Temporary repair work on bridge.

The code in question had similar clauses to those of *BS 8110*[1] that provide an expression that increases the shear resistance according to how much moment is required to nullify the effects of prestress. The following example comparing calculations in *BS 8110*[1] and *BS EN 1992 (EC2)*[2] shows how such an anomaly can occur.

Design information:

Shear force V = 9000 kN
M_0/M = 0.4
Prestress = 0.8 MPa
f_{pe}/f_{pu} = 0.5
d = 2700 mm
b = 800 mm
f_{ck} = 30 MPa
f_{cu} = 37 MPa
f_y = 500 MPa

Main reinforcement: 12 bars (32 mm diameter) to simulate both prestressing tendons and reinforcement
Links: 16 mm diameter (six per section) at 300 mm spacing

BS 8110[1]: *(Cl. 3.4.5.4)* $v_c = 0.79\{100A_s/(bd)(f_{cu}/25)\}^{1/3}(400/d)^{1/4}/1.25 = 0.55$ MPa
(where $(400/d)^{1/4} \geq 1$)
$V_c = bdv_c/1000$ = 1189 kN
(Cl. 4.3.8.5) $V_{cr} = (1 - 0.55f_{pe}/f_{pu})\ b\ d\ v_c + V\ M_0/M = 4462$ kN

V_{cr} and V_c are the concrete shear resistances without shear reinforcement with and without prestress. The large difference, the shear resistance from prestress, 3255kN, is questioned by some.

The shear resistance of the shear reinforcement was

$$V_s = (A_{sv}/s)\ 0.87f_y d \qquad\qquad = 4723\ kN$$

The total shear resistance is thus

$$V_t = V_{cr} + V_s \qquad\qquad = 9185\ kN$$
$$\text{OK}$$

EC2[2]: The calculation of the concrete shear resistance without shear reinforcement, $V_{Rd,c}$, is similar to that of V_{cr} in *BS 8110*[1], except that it includes the effects of prestress by factoring the axial prestress.

(Cl. 6.2.2) $V_{Rd,c} = [0.12(1 + \sqrt{(200/d)})\ (f_{ck}\ 100A_s/bd)^{1/3} + 0.15\sigma_{cp}]bd/1000 = 1042\ kN$

The reduction factor for this section compared with a section 200 mm deep is 0.64.

The calculation for shear resistance with shear reinforcement uses the variable truss method. Choosing cot θ = 2 for this example the shear resistance of the links is

$$\textit{(Cl. 6.2.3)}\ V_{Rd,s} = (Asv/s)\ z\ f_{ywd}\ \cot θ/1000 \qquad = 8497\ kN$$
$$\text{Not OK}$$

and the shear resistance of the concrete strut is

$$V_{Rd,max} = α_{cw}\ b\ z\ ν_1\ f_{cd}\ /\{1000(\cot θ + \tan θ)\} \qquad = 8540\ kN$$
$$\text{Not OK}$$

The total shear resistance permitted by *BS 8110*[1] is the sum of the resistances of the concrete and shear reinforcement. No reduction is required for the deep section. This is in contrast to calculation to *EC2*[2] which a) reduces the shear resistance of the concrete without shear reinforcement by a large factor and b) only permits the resistance of the shear reinforcement or concrete strut.

Comment — Although this does not represent a very unsafe situation, it does emphasize the importance of the increased reduction factor for deep sections in *EC2*[2].

This example also demonstrates the very different approaches of the two codes of practice to shear resistance and especially shear resistance of prestressed sections.

Chapter 4

Failures due to Misuse
of Code of Practice Clauses

4.1 FLAT SLAB AND TWO-WAY SLAB BEHAVIOUR

A flat slab is defined as a plate supported on individual columns. A two-way slab is a slab supported by beams at each edge. The UK codes of practice differentiate between flat slabs and two-way slabs.

A car park was designed (based on a design–build contract) as a coffered slab supported by columns. The coffers were omitted around the columns so that the solid section in this region provided punching shear resistance. The designer clearly understood that the system would behave as a flat slab but decided to use the simplified applied moment coefficients for a two-way slab, ignoring any beam effects.

The maximum moments for a two-way slab, taken from Table 3.14 of *BS 8110*,[1] are:

Hogging $0.031\ nl^2$
Sagging $0.024\ nl^2$

The maximum moments for a flat slab, taken from Table 3.12 of *BS 8110* are:

Hogging and sagging for interior spans $0.063\ Fl$ or $0.063\ nl^2$
Hogging and sagging for outer span $0.086\ Fl$ or $0.086\ nl^2$

The reinforcement detailed was thus less than half that required for a flat slab! For some reason, the contractor fitted only half the reinforcement detailed for the supports. The slab finished up with only a quarter of the required reinforcement in some areas.

After 10 years of service, cracks were appearing in the slab and a decision had to be made to determine what remedial work was required. Although the car park did not appear to be about to collapse, the remedial work required to ensure its safety was considered too extensive and the structure was demolished.

Comment — This was a clear example of a design error compounded by further errors on site. One reason that the structure did not collapse was that the actual loading was much less than the design value of 1.5 kN/m². The actual load on a car park may be as little as half this value. Another reason for the apparent strength is likely to be membrane action effects not included in the design.

This is an example of the benefits of an indeterminate structure. There existed alternative load paths to that considered in design, which prevented collapse of the structure.

4.2 RIBBED SLAB SUPPORTED ON BROAD BEAM

Several buildings of a university included ribbed floor construction (see Figure 4.1). The ribbed slab was designed as a simply supported element spanning between edges of a broad beam. The curtailment of the longitudinal reinforcement in the ribs was in accordance with the code clause stating that, 'each tension bar should be anchored by one of the following: … b) an effective anchorage length equivalent to 12 times the bar size plus d/2 from the face of the support….' In this case, the curtailment was measured from the edge of the broad beam. Top reinforcement in the slab was provided in the form of a structural fabric.

The depth of the broad beam was the same as the trough slab and the design wrongly assumed that it could be considered a one-way slab and therefore did not require links (see Figure 4.2).

In order for the edge of the beam to be considered in the design as the support face, the vertical force from the load on the ribbed slab had to be

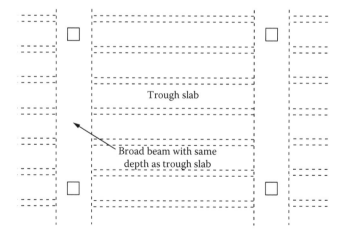

Figure 4.1 **Plan view of trough slab.**

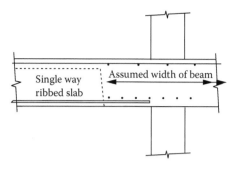

Figure 4.2 **Section through broad beam.**

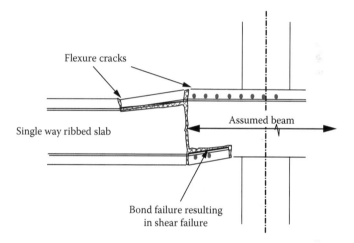

Figure 4.3 **Section through ribbed slab at failure.**

transferred to the top of the beam. Vertical reinforcement should have been provided for this. Links in the broad beam (in addition to any links necessary for shear resistance) could have provided sufficient resistance.

Failure of the ribbed slab occurred at the support with the broad beam (see Figure 4.3). Complete collapse was avoided by immediate temporary propping. The sequence of the failure was likely to be:

- Yielding of the fabric reinforcement in the top flange of the slab
- Large flexural cracks opening near the support
- Redistribution of moment transferred tension stresses to the bottom reinforcement
- Combination of anchorage and shear failure

Comment — Failure could have been avoided if (1) the span of the ribbed slab had been taken as the centre-to-centre distance of the broad beams, or (2) links had been provided in the broad beam. A better design would have included both features.

4.3 CAR PARK COLUMNS

Figure 4.4 shows the beams on the sides of the internal columns. They are stepped to allow ramping between levels (as is often the case). The internal columns were much stiffer than the edge columns and in consequence the support moments and shear forces within the column were high compared with the edge columns. The column shear forces caused large shear cracks (up to 2 mm) within many of the columns.

At first sight, this might be considered as an ultimate limit state. However there was still a load path for the vertical forces through the column and when the columns cracked, the moment in the adjacent beam was redistributed to the span. Fortunately, the span resistance was adequate.

Although a failure mechanism was not present, a problem occurred from repairing the cracks. As soon as the car park filled with vehicles after the repair, the cracks reopened and eventually the deterioration of the column concrete could have led to a collapse.

The design had been carried out to the existing code of practice of the time which did not have specific clauses for shear in columns. By the time remedial work was carried out, new code clauses in place required shear checks of columns. The design of this particular joint was inadequate to the new clauses. The insurance company required remedial work to be carried out so that the building would comply. This was considered to be so costly that the decision was made to demolish the existing structure and replace it with a new design.

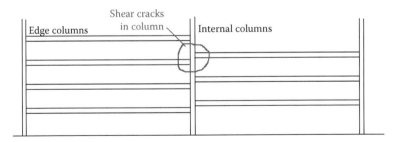

Figure 4.4 Structural elevations of beams and columns of car park.

Comment — Although the existing design code did not cover this particular situation, the designer should have been aware of the problem and taken action in the design to prevent overstressing within the joint. This is an example of common practice in the 1960s and 1970s when many engineers extrapolated the existing clauses of the codes of practice for situations outside the scopes of the codes.

Typically continuous beams were designed assuming moments were not transferred to columns. This ensured safe designs for the beams but meant that columns could become overstressed, as occurred in this situation.

Chapter 5

Problems and Failures due to Inadequate Assessment of Critical Force Paths

5.1 HEAVILY LOADED NIBS

It is possible that a heavily loaded nib may require more than one load path to transfer a load safely. Strut-and-tie models can demonstrate alternative methods for reinforcing. However, it should be realised that the least direct paths will cause the most distortion and cracking, and should not be used for the serviceability state.

Figure 5.1 shows the most direct strut-and-tie (primary) model. The force paths are closest to that of an elastic model and will create the least internal distortion to achieve equilibrium. Figure 5.2 shows a secondary strut-and-tie model. This may be accompanied by distortion and cracking of the concrete before it can achieve equilibrium.

If the forces on the nib are too great for the primary model, it is reasonable to superimpose the secondary model to provide sufficient resistance for the total ultimate loads. However, it is important to ensure that the primary model is sufficient to resist the serviceability loads and provide crack control.

Comment — This approach was used for a major viaduct. The combination of the two models (see Figure 5.3) enabled all the reinforcement to fit—just!

5.2 SHEAR WALL WITH HOLES AND CORNER SUPPORTS

A multi-storey shear wall required so many openings (windows, doors, etc.) that the load path became very complicated. The designer assumed that the load would flow to the corners and then track vertically down the edge of the wall (see Figure 5.4a). In fact, since the wall was built in situ as a homogeneous structure, strain compatibility caused the load to flow back into the full width of the wall. The result was that several storeys of load

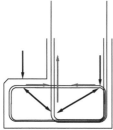

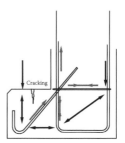

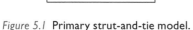

Figure 5.1 **Primary strut-and-tie model.**   *Figure 5.2* **Secondary strut-and-tie model.**

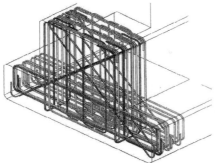

Figure 5.3 **Example of use of the two strut-and-tie models. (Courtesy of Gill Brazier.)**

were supported by a deep beam that transferred the load to its end supports (see Figure 5.4b).

The limiting height of the natural arch of a deep beam (0.6 × span) was not considered (see Figure 5.4b) and this resulted in the omission in the design of much of the reinforcement needed for the bottom tie. Construction had reached several floors up by the time the mistake was recognised and this led to a redesign of the wall during construction and heavy remedial work. Each part of the wall required careful re-appraisal. This led to the requirement of much more reinforcement at each floor level. The bottom corner reinforcement details required special attention to ensure that the junction between the tie and compression struts was adequately designed.

Figure 5.5 shows in a simple diagrammatic form how the force paths automatically flow out and back again. The assumed force path down the edges would not require ties at top and bottom, but without these the actual force path would cause large cracks to open up from the top and bottom surfaces. Even after cracking, the angle struts would still exist and so would the consequential horizontal component. Without sufficient tie force to resist, the support joint would move outward and eventually failure would follow.

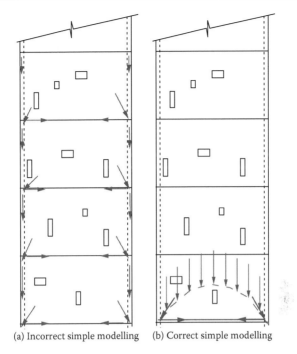

(a) Incorrect simple modelling   (b) Correct simple modelling

Figure 5.4 **Multi-storey shear wall.**

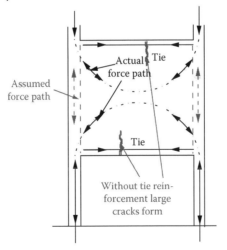

Figure 5.5 **Modelling deep beams.**

Comment — The consequence of missing this simple principle of deep beam behaviour before construction reached such an advanced state meant that it required the redesign of the structure and reprogramming of construction which were extremely costly.

5.3 DESIGN OF BOOT NIBS

Where nibs are attached to the bottom of a beam it is important to understand the load path of the forces. Figure 5.6 shows a typical section of such a nib.

The conventional assumption for a short cantilever of d_c and z_c (shown in red in Figure 5.6) is unsafe for such a nib. The design compression zone for such a model would be close to the bottom face of the beam and likely to fall outside the beam reinforcement (both the links and main reinforcement). Strut-and-tie modelling is helpful to explain why this is so. The strut (shown in red in Figure 5.6) would just cause the cover to the reinforcement to spall off. The strut must be supported mechanically by the reinforcement of the supporting beam (shown in black in Figure 5.6). The effective lever arm becomes much smaller and the tension force in the nib top reinforcement much larger than assumed by the short cantilever approach.

It should also be noted that the force in the supporting links of the beam, F_{t2d}, is likely to be much greater than the applied load on the nib, F_{Ed}, to satisfy equilibrium. For the situation shown in Figure 5.6, it is conservative to assume the compression acts at the centroid of a triangular compression stress block. Hence the force in the link, in addition to any shear, may be calculated as follows:

$$F_c = F_{Ed} \times a_c/z_b$$

$$F_{t2d} = F_c + F_{Ed} = F_{Ed}\,(1 + a_c/z_b)$$

Comment — There are probably many nibs of this type that have been designed incorrectly and survive because of built-in safety factors and the fact that the load assumed in the design has not occurred.

The error described in this case study was found in the design of a nib for a very prestigious project. It was very fortunate that it was discovered before construction started.

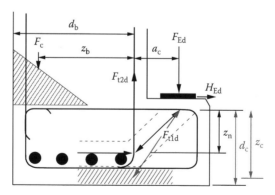

Figure 5.6 Nib attached to bottom of a beam.

Chapter 6

Problems and Failures due to Poor Detailing

Poor detailing is often connected with a lack of sufficient design thought and accompanied by poor workmanship in construction. The combination can lead to structural failure. Many of the case studies in this chapter have resulted from extrapolations from previous jobs. Small modifications were made to save construction time and cost. This was not accompanied by sufficient checks to ensure a safe structure.

6.1 CONCRETE OFFSHORE PLATFORM

The platform included a large cellular concrete structure below the three towers as shown in Figures 6.1 and 6.2. During construction, the platform underwent submerging for deck mating after which the plan was to raise it again and tow it to its final position in the oil field. It was during the submerging prior to deck mating that one of the tri-cells failed. This caused flooding of the structure and further uncontrolled sinking that led to an implosion of the structure and complete collapse.

Figure 6.3 shows a detail of the tri-cell wall that was designed to resist the water pressure when the cellular base was submerged. Figure 6.3b shows the original form of the cells with cylindrically shaped walls. The natural arch action provided by that geometry was not present in the modified form shown in Figure 6.3a.

The analysis of the cell structure was carried out using a finite element software package. The accuracy of the analysis was reduced as the arrangement of the quadrilateral elements meant that those in the region of the tri-cell corners were distorted from the ideal square shape. This led to errors in the results. It was realised after the failure that the shear stress results from the analysis were in error on the unsafe side.

The critical shear section was reinforced with T-headed bars. The software package gave the required areas of reinforcement, and the design required that the length of the T-headed bars would extend across the full

Figure 6.1   Concrete offshore platform during construction. (Courtesy of Gillian Whittle.)

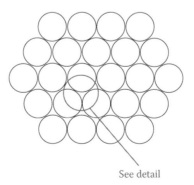

See detail

Figure 6.2 **Plan section of cell structure.**

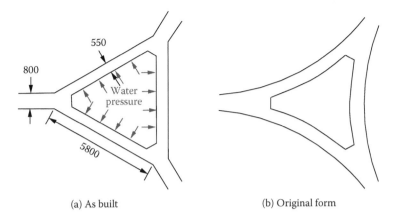

(a) As built   (b) Original form

Figure 6.3 Detail of tri-cell.

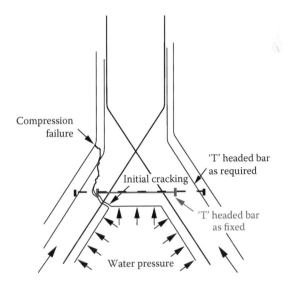

Figure 6.4 Section through tri-cell where failure occurred.

width of the section. As they were difficult to fix through the outer layer of reinforcement, it was decided to reduce their length (see Figure 6.4).

A crack formed at a corner of the cell and spread to the end of the T bar. The water pressure became active in the crack, making the situation worse. A shear crack developed up to the compression zone of the section and this failed in a brittle manner. The resulting massive leak led to the structure sinking further until the increased water pressure caused progressive failure of the whole cellular structure. It finished up on the sea bed as a mass of rubble.

Comment — This catastrophic failure was the result of a number of errors:

- The analysis program was set up with a finite element mesh that was too coarse to provide accurate shear results.
- The T-headed bars were too short and allowed the shear resistance to become unsafe. This was probably the primary cause of the failure.
- There was minimal checking of the design and detailing, possibly because this was a type of structure that was well established. In the same period, the same designer was involved in three quite different and more complicated platforms.
- In previous designs, the geometry of tri-cells had been formed by intersecting cylinders. Although some cracking occurred, the geometry ensured that sufficient arching action took place without inducing large shear stresses. The geometry of tri-cells was altered on this project to make the formwork simpler to construct. Unfortunately the new form did not allow arching action to take place and the sharp corners acted as crack inducers (these were also present in the original design).
- The rebuild retained the cylindrical geometry in the tri-cells and the reinforcement was detailed to ensure mechanical linkage. The T-headed bars were extended to the outer reinforcement.

6.2 ASSEMBLY HALL ROOF

This disaster could also be called *the miracle of the decade*. On 13 June, 1973, late in the evening, the roof of an assembly hall crashed to the ground. In the words of the caretaker, he heard a loud rumble, went to investigate by torch light, and found the whole roof weighing many tons had collapsed (see Figures 6.5 to 6.7). Twenty-four hours before this event, *some five hundred parents had attended a meeting in the hall* and the chairs were still in place.

The principal cause of the collapse was inadequate bearing for beam seatings and deterioration of concrete at beam ends. The hall was one of the first buildings found to have suffered from the effects of high alumina cement (HAC; see Section 7.4).

Comment — This was an example of inadequate design and poor detailing of the end bearing nibs built into the supporting beam for the precast beams. The reduction in strength caused by HAC left no margin for temperature effects. The combination was likely to have triggered the collapse.

Bearings for precast beams

Figure 6.5 Assembly hall showing edge beam that supported the precast beams. (From Scott, G.A., *Building Disasters and Failures*, Construction Press Ltd., Lancaster, 1976. With permission.)

Figure 6.6 Part of roof that collapsed on chairs below. (From Scott, G.A., *Building Disasters and Failures*, Construction Press Ltd., Lancaster, 1976. With permission.)

Figure 6.7 Collapsed roof lying on floor below. (From Scott, G.A., *Building Disasters and Failures*, Construction Press Ltd., Lancaster, 1976. With permission.)

6.3 UNIVERSITY BUILDING ROOF

On a June morning, following a period of cool nights and very hot sunny days, a portion of a roof, including three of the prestressed beams, collapsed on to the floor below. A fourth beam had pulled out of its seating but remained in place jammed against the edge beam.

The roof was constructed of prestressed precast concrete beams made with HAC spanning 12.6 m, supporting precast concrete slabs acting as permanent formwork for an in situ concrete topping (see Figure 6.8). The outer ends of these beams rested in 50 mm deep pockets cast into the vertical inside faces of the edge beams. In addition to these, the edge beams had recesses just below their seatings. These encroachments on the already narrow section led to the adoption of a reinforcement arrangement that had no main steel under the seating pockets for the prestressed beams.

The cause of this collapse was a disastrous combination. The columns and edge beams were precast as T units (with webs facing outward). The architect had, in order to obtain a dark colour, specified HAC concrete to be made with aggregate that was highly alkaline. The specification had been mistyped or misread to call for one part HAC to four parts sand to two parts aggregate (these mix proportions were confirmed by analysis after the event). Cores drilled from one of the T units after the collapse revealed strengths as low as 7 MPa and some could not be extracted in one piece.

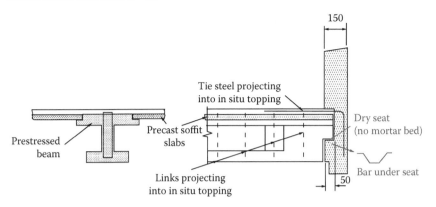

Figure 6.8 **Prestressed beam and support.**

The depth of seating for the pretensioned roof beams satisfied the code requirement for bearing pressure, but left no margin for erection tolerances or temperature effects. There was inadequate reinforcement detailed under the beam seatings, and as the T units were cast with external faces down, even that reinforcement was displaced downward, away from the seating ledge.

Bearing stresses on the shallow seating and vertical stresses under the pockets were within the code recommendations, and only an inverted hat bar was provided under the beam seating; this was detailed in a way that made it difficult to construct in the correct position as the edge precast beams were cast with the outsides face down to produce a smooth fascia.

Links projected from the prestressed beams into the in situ topping and the edge beams had 9.5 mm mild steel bars projecting into the in situ topping, thus providing some tying together, but *not* at the level of the seating.

The actual mechanism that led to the failure was found to be thermal hogging of the prestressed beams. The tie steel projecting from the edge beam into the in situ topping acted as a hinge, and the end rotation accompanying the thermal flexure between the two extremes corresponded to a horizontal movement at the level of the seating of approximately 2 mm. This caused cracking of the bearing nib as shown in Figure 6.9. The tension stress in the concrete support reduced its shear resistance sufficiently to cause the failure. The prestressed beam had been made with high alumina cement and although it aggravated the situation it was *not* the cause of the failure.

On another building, an identical detail was found, but the edge beams were constructed with ordinary Portland cement concrete. In that case, a crack similar to the one shown in Figure 6.9 was found, but presumably the reinforcement had not been badly displaced, as full collapse had not occurred. This did, however, demonstrate that HAC was not the primary cause of the collapse of this roof, but accelerated the cracking until collapse had become inevitable.

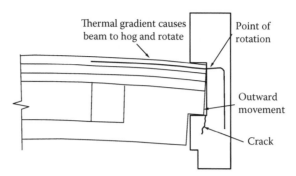

Figure 6.9 **Detail at support.**

Comment — In summation, the collapse was caused by:

- Unfortunate selection of aggregate and ignorance of the material's properties
- Poor detailing of the edge joint
- Insufficient bearing in the edge beam
- Insufficient horizontal restraint at the level of the bearing to resist the relative movement due to temperature effects
- The reinforcement detailed under the bearing was displaced

6.4 MINIMUM REINFORCEMENT AND CRACKING

The design of parts of a bridge deck slab did not require more than minimum reinforcement. Some months after construction large cracks began to appear. The sizes of some cracks increased to 2 mm and the reason was not immediately clear.

One requisite of a code of practice for crack control is to satisfy the minimum reinforcement percentage. Unless this is provided, the spacing and sizing of bars have little effect on crack width. The reason for this is based on the tension strength of the concrete. If for any reason (shrinkage etc.) the concrete cracks, it is essential that the reinforcement does not yield at the crack. If it does yield, the crack will become large and this will prevent the occurrence of small cracks at small spacings. In order to avoid such a situation, the tension yield strength of the reinforcement should be at least equal to the tension strength of the concrete. Codes of practice stipulate a minimum amount of reinforcement based on the tension strength of the concrete.

The design for this project specified a concrete strength of $f_{ck} = 30$ MPa and the amount of minimum reinforcement percentage was calculated and provided for this value. Unfortunately, the contractor provided concrete with a strength of 50 MPa. Because of its higher tensile strength (4.1 MPa compared with 2.9MPa), the reinforcement yielded when the concrete cracked and the cracks became large—up to 2 mm!

Comment — This is one of the few, but not insignificant, cases where increasing a material's strength makes the situation worse.

6.5 PRECAST CONCRETE PANEL BUILDING

In the early hours of 16 May, 1968, a gas explosion in a bathroom on an upper floor shook a building, resulting in the instantaneous collapse of part of one wing (see Figure 6.10). Four people were killed. Figure 6.11 shows a closer view of the upper floors after the explosion.

Figure 6.10 Progressive collapse showing point of explosion. (Ronan Point building collapse, May 16, 1968, London. Courtesy of Building Research Establishment.)

Figure 6.11 Closer view of upper floors showing point of explosion. (Ronan Point building collapse, May 16, 1968, London. Courtesy of Building Research Establishment.)

The reasons for the collapse were:

1. The possibility of unusual, and hence non-codified, loads was not considered.
2. The structure was inadequately tied together.

Traditional pre-World War II two-storey housing would not have had any engineering input; brick wall thicknesses and timber floor joist sizes were prescribed by the London City Council Building by-laws, and similar regulations outside London. There had been gas explosions before this incident in similar types of dwellings, but the damage and casualties had usually been limited to one household. The risk was accepted as a 'fact of life.'

There were therefore no precedents for progressive collapse, when system building was introduced. For four-storey walk-up blocks, the materials and thicknesses of load-bearing walls would similarly be prescribed. Any fire-breaking floors would be straightforward reinforced concrete slabs, designed for occupational loading. They were cast in situ onto the walls below and therefore 'stuck to the walls.' Even if the design span was parallel to the wall below, the slab would impose some load on it. That, together with the load from the wall above, would provide so much preload that a lower level wall would be unlikely to be blown out due to its prescribed thickness.

There was therefore a degree of tying together, albeit reliant on friction, in traditional construction. In large-panel construction, most of this inherent tying together was lost and not replaced (with steel ties) until the requirement for resistance against progressive collapse entered the regulations and codes.

Dry packing of the panel joints, whether with mortar or fine concrete, is carried out as a menial task after the 'spectacular' event of placing and plumbing the panel. It is likely to attract less careful workmanship and supervision.

Comment — This collapse was a significant event for the industry in the UK and marked the partial demise of the precast industry. Large precast panel and frame construction became much less popular in the following two decades. Information gathered from the incident led to major changes to the UK's *Building Regulations* (1970)[11] and codes of practice (starting with the *CP 116*[12] precast concrete code in 1970) with regard to progressive collapse and robustness. More recently, the Eurocodes have included accidental load and robustness clauses.

6.6 FOOTBRIDGE

A footbridge was being constructed over a motorway. The bridge had two spans with an in situ concrete T-section deck (see Figure 6.12). The span lengths were 20 and 40 m. After removing the formwork and props, two large transverse cracks (2 and 3 mm) appeared in the sides and soffit of the longer span. The cracks occurred at the positions of the laps of the main bottom reinforcement.

The reinforcement had been detailed so that all the main bars in the bottom, 16 T40s, were lapped at positions of high stress. It is generally considered poor detailing practice to lap bars in positions of high stress and lapping

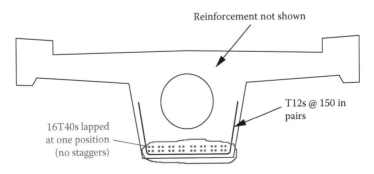

Figure 6.12 **Section through footbridge.**

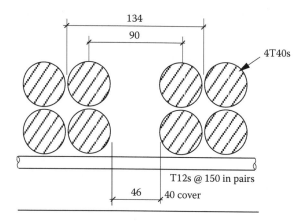

Figure 6.13 Layout of bars at laps.

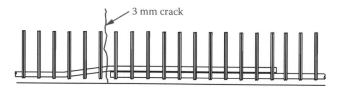

Figure 6.14 Side view of bottom reinforcement showing position of crack.

all the bars in a congested situation compounds the error. The bars were cranked to allow them to be as close as possible (see Figures 6.13 and 6.14). Links, T12s @ 150 in pairs enclosed all the main bars. No extra links were provided at the laps or cranks.

The lap joints should have included vertical links surrounding each pair of lapped bars. The presence of the vertical cranked bars increased the force to be resisted at that end of the lap and additional links should have been provided at this position. The force due to the cranked bars acted as a crack inducer.

Cores were taken and X-ray examinations showed that the laps of the main reinforcement had been detailed as shown on the drawings and that construction had followed the drawings. It was decided that the bridge should be demolished.

Comment — This failure was a result of poor detailing. Lapping of T40 bars creates large transverse forces that must be resisted with links. If the code rules (*BS 8110*[1]) had been applied, much more transverse reinforcement should have been specified. Good detailing practice would have avoided lapping the bars at the position of maximum stress.

Chapter 7

Problems and Failures due to Inadequate Understanding of Materials' Properties

7.1 CHANGES OVER TIME

Over the years, the changes in materials and reductions in safety factors make it more important to understand the behaviour of reinforced concrete and provide more care. Rules of thumb and empirical methods may have been developed for different conditions and may not be applicable for today's materials' properties and design criteria need to be checked to determine whether they are still applicable. Another result of the development of and changes to material properties is that the ultimate limit state is often no longer critical, and a design now often depends on the serviceability limit states, apart from punching shear.

Concrete — There has been a continuous increase in the strength of concrete over the last hundred years; much of the increase has developed since 1980 (see Figure 7.1). Around that time, the value of blended cements and the use of admixtures was realised. Modern concretes have become complex with almost infinite variations available depending on the requirements. The understanding of how to change the properties of concrete and reinforcement is developing rapidly. It includes:

- The use of admixtures and blended cements. Admixtures are essential for modern concrete. Self-compacting concrete is one important example. Blended cements allow the control of the rate of strength gain and the amount of heat created.
- The use of stainless steel will increase for situations where durability is paramount.
- The use of higher strength concrete will become more popular for floor slabs, particularly flat slabs. This will result in thinner and longer span slabs.
- Serviceability limit states have already become critical to flat slab design and it will become more common to check vibration of floors.
- The use of fibres will increase; the use of steel fibres has already been proven for ground floor slabs.

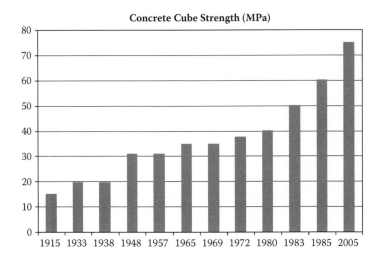

Figure 7.1 Increase of concrete strength during 20th century.

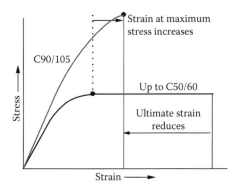

Figure 7.2 Change in stress block for high strength concrete.

All these developments add complexity and cost to concrete construction. Above a cylinder strength of 50 MPa, the stress–strain properties change with increases in strength. Figure 7.2 shows this change diagrammatically. The concrete itself becomes more brittle as the strength increases, but it should be noted that in flexural members (beams and slabs), the ductility and brittleness are dependent mostly on the properties of the reinforcement.

The increase in concrete strength and reduction in overall factor of safety (see Figure 7.3) have meant that, for many structural elements, the design for the serviceability limit state is becoming more critical than that for the ultimate limit state.

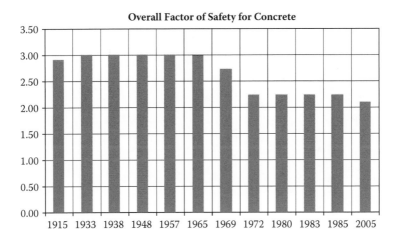

Figure 7.3 Reduction in concrete partial safety factor during 20th century.

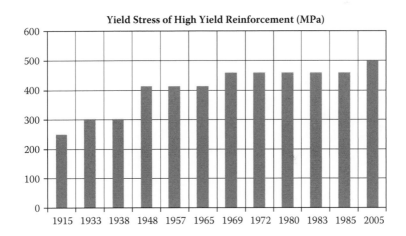

Figure 7.4 Increase in steel yield strength during 20th century.

Reinforcement — A similar pattern of change has occurred for reinforcement both in strength and partial safety factors (see Figures 7.4 and 7.5).

7.2 REBENDING OF REINFORCEMENT

In 1964, the construction of a 35 m high dust bunker for a coal-fired power station included an external concrete cantilever staircase to be built on to the face of the outside wall of the bunker. The construction of the wall

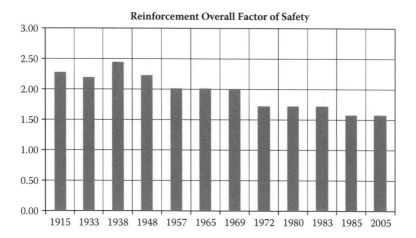

Figure 7.5 Reduction in reinforcement overall safety factor during 20th century.

meant that the reinforcement required for the stairway would be cast flush with the wall and then bent out after removal of the formwork. At that time, proprietary reinforcement systems for such a situation did not exist and the bars were bent before fixing within the shutter. The radius of bend would have been to a standard of three times the bar diameter.

After the formwork had been removed, the surface of the concrete was scabbled to expose these bars and the scaffold tubes were threaded over them. The scaffold tubes were then used to lever the bars out of the wall into their final positions. About 30% of all the bars bent out snapped off during the operation. The reason was a combination of factors:

- The bars should have been bent out with a special tool that ensured that the radius of bend was at least three times the bar diameter.
- The particular batch of reinforcement was found to be more brittle (less ductile) than specified.
- The work was carried out at a temperature just above freezing.

The remedial action taken was to drill holes into the concrete and grout in replacement bars.

Comment — The bending out of reinforcement cast into walls is a common procedure and, all too often is done with scaffold tubes that are readily accessible on site. It is regrettable that a proper rebending tool is not often used which is a reflection of poor understanding of the material's physical and chemical properties. In the manufacture of reinforcement, special procedures are in place to check the rebending of bars to ensure that the reinforcement is sufficiently ductile. It is unfortunate that some manufacturers have continually tried to eliminate such tests from the reinforcement standard.

7.3 TACK WELDING OF REINFORCEMENT

The design of a building with large columns required 32 mm diameter starter bars projecting from the pile caps. The temporary works for construction included tack welding some small diameter bars to the starter bars. When the time came to fix the column reinforcement to the starter bars, the contractor attempted to bend the starter bars to ensure that they would fit into the column shutter with sufficient cover to the concrete face. A large sledge hammer was used to effect this. During this operation, two of the 32 mm diameter bars snapped off.

The reason was that tack welding the small bars onto the larger diameter starter bars changed the molecular structure of the latter. Unlike structural welding, tack welding heats just the local spot, and the heat sink of the main bar cools it very rapidly. The result was that the starter bars became brittle and required only a sharp blow to fail. In the past, tack welding on site was forbidden. Today it is sometimes permitted if carried out by a skilled specialist. Unfortunately once permitted, it is all too easy for a non-specialist to do this work, believing that it will do no harm.

Comment — Too many people are unaware that tack welding can have significant structural effects. This is another case where a material's chemical and physical behaviour was not properly understood.

7.4 HIGH ALUMINA CEMENT

High alumina cement concrete has achieved a certain notoriety following the collapse of several buildings in the 1970s. By the end of 1974, up to 50,000 buildings had been reported as suspect and a major effort was made to check their safety. Fortunately, many of the affected beams stood in dry conditions and the chemical deterioration had not reached an advanced stage. The worst affected elements were positioned in damp environments.

Description — High alumina cement is manufactured from limestone or chalk and bauxite (the ore from which aluminium is obtained). The two materials are crushed and fired together using pulverised coal as a fuel. The materials fuse together, and after cooling are crushed and ground into a dark grey powder.

The predominant compounds are calcium aluminates; calcium silicates account for no more than a few percent. The calcium aluminates react with water and the primary product is calcium aluminate decahydrate (CAH_{10}).

One of its main characteristics is that the concrete made with it achieves its full strength after 24 hours compared with 28 days for a concrete with Portland cement. However, its crystal structure is unstable and changes to tricalcium aluminate hexahydrate (C_3AH_6) spontaneously (albeit slowly). This process occurs at room temperature and is accelerated by an increase in temperature. The crystal structure transforms itself to a more compact form, with the result that the cement matrix of the concrete becomes porous and weaker. The extent to which this conversion, as it is known, occurs is largely a function of the:

- Original water/cement ratio of the concrete
- Temperature rise in the concrete during hardening
- Temperature and moisture to which the hardened concrete is subsequently exposed

Degree of conversion — It was found that at the time of the collapses most HAC concrete used in buildings was 90% or more converted. A concrete from a wet mix exposed subsequently to the sun was found to have its strength reduced from 40 MPa at 24 hours to an average of about 10 MPa after less than 10 years. In contrast, concrete from prestressed precast beams with a low water-to-cement ratio and hence a 24 hour strength of 65 MPa from the same building but in a dry environment was found to have retained a strength of about 35 MPa.

7.5 CALCIUM CHLORIDE

Many reinforced concrete structures have suffered from too much chloride in the concrete mix. This causes the breakdown of the high level of alkalinity. When moisture and oxygen are present, carbonation occurs. This allows the reinforcement to rust and leads to spalling of the concrete surface.

Before 1980, calcium chloride was used extensively for in situ concrete works, frequently without adequate supervision. It was used principally for frost protection and to facilitate the rapid stripping of shutters. However, all too often, too much was added. In the 1980s, the codes of practice and concrete specifications were tightened to ensure that the rusting and spalling should not happen again. The following three examples describe where too much chloride in concrete caused structural failures.

Example 1 — A primary school (built in 1952) was shut in 1973 due to extensive corrosion of the reinforcement of factory-made precast concrete beams. This was due to the presence of too much calcium chloride added during the manufacture of the beams to hasten the hardening of the cement. The condensation under the beams accelerated the corrosion by combining with the calcium chloride to produce hydrochloric acid.

Example 2 — In 1974, the concrete roof of a school collapsed. The reason was found to be too much calcium chloride in the concrete, causing the reinforcement to deteriorate and eventually fail.

Example 3 — An independent investigation of the collapse of a 100 m long pedestrian bridge found the cause to be high levels of calcium chloride in the grout used in the ducts for the prestressed tendons. This led to corrosion and failure of the prestressing tendons.

7.6 ALKALI–SILICA REACTION

The alkali–silica reaction (ASR) is a heterogeneous chemical reaction that takes place in aggregate particles between the alkaline pore solution of cement paste and silica in the aggregate particles. Hydroxyl ions penetrate the surface regions of the aggregate and break the silicon–oxygen bonds. Positive sodium, potassium, and calcium ions in the pore liquid follow the hydroxyl ions so that electro-neutrality is maintained. Water is imbibed into the reaction sites and eventually an alkali–calcium–silica gel is formed.

The reaction products occupy more space than the original silica so the surface reaction sites are put under pressure. The surface pressure is balanced by tensile stresses in the centres of the aggregate particles and in the ambient cement paste. At a certain point, the tensile stresses may exceed the tensile strength and brittle cracks propagate. The cracks radiate from the interior of the aggregate out into the surrounding paste.

The cracks are empty (not gel-filled) when formed. Small or large amounts of gel may subsequently exude into the cracks. Small particles may undergo complete reaction without cracking. Formation of the alkali–silica gel does not cause expansion of the aggregate. Observation of gel in concrete is therefore no indication that the aggregate or concrete will crack. ASR is diagnosed primarily by four main features

- Presence of alkali–silica reactive aggregates
- Crack pattern (often appearing as three-pointed star cracks)
- Presence of alkali–silica gel in cracks and/or voids
- $Ca(OH)_2$ depleted paste

In mainly unidirectional reinforced members, the cracks become linear and parallel to the reinforcement. The degree of cracking depends on the amount of confining reinforcement, i.e., links, etc. One major concern was that ASR caused cracking that led bits of concrete to fall off structural elements and hit people below. This led to demolition of the structures in some cases. Examples of ASR effects are given in Figure 7.6.

Figure 7.6 Examples of alkali–silica reaction. (Top: From the US Department of Transportation Highway Administration; middle: From Dr. Ideker, http://web.engr.oregonstate.edu/~idekerj/; bottom: From the US Department of Transportation Highway Administration.)

7.7 LIGHTWEIGHT AGGREGATE CONCRETE

During the 1960s, a medium-size civil engineering contractor wanted to join the housing drive, then at its peak. At the time, an Austrian construction firm used crushed brick rubble as aggregate in un-reinforced concrete walls for six- and seven-storey blocks of flats. Inspired by this, it was decided to try to develop a similar form of load-bearing wall with adequate thermal insulation, made of lean-mix plain concrete with light expanded clay aggregate (LECA). A 12-storey block was constructed as a pilot project.

The strength of the wall concrete was reduced in stages—about 2000 psi (14 MPa) at 28 days for the four bottom storeys, 1600 psi (15 MPa) for the next four, and 1200 psi (8 MPa) for the top storeys. The floor slabs were of traditional reinforced concrete, but the roof slab was reinforced LECA concrete with a strength of 3000 psi (21 MPa).

There was no significant adverse feedback from the tenants nor the building authority. The block remained standing and in use for over 40 years. Encouraged by the apparent success, the contractor started promoting the 'system.' About the same time, lightweight aggregate concrete was included in the code of practice and a minimum strength of 3000 psi (21 MPa) was stipulated. This required a richer mix than that used for the walls of the earlier block. The resulting effects of this on the thermal insulation and shrinkage properties of the LECA concrete appear to have been overlooked by the design team.

A few blocks were built for local authorities outside the London County Council area. These were higher than the first block utilising the higher concrete strength required by the code in the walls. Many of the flats were allocated to tenants in poor financial circumstances, who could not afford the charges for the underfloor heating and used paraffin heaters instead. This, combined with the reduced thermal insulation of the external walls, led to severe condensation.

Structurally more important, however, were the diagonal cracks that developed on the top floor of one of the blocks within a short time after hand-over. From their geometry, they appeared to be due to lower shrinkage and greater thermal expansion of the roof slab relative to the wall concrete.

Definitely alarming was the occurrence of horizontal cracks in one of the 200 mm thick internal cross walls connected to the 300 mm thick external wall at right angles. One of these cracks on the 13th floor of a 16-storey block opened suddenly with a noise like a gun shot. The wall was 200 mm thick and, according to the design assumptions, carried the floor slabs that spanned about 3.5 m on either side. This meant that the building contained three storeys of unreinforced concrete cross wall, with the load from approximately 3.5 m width of floors plus the roof hanging or cantilevering off the external wall!

Structurally, the only explanation for these cracks seemed to be that the internal wall was drying out, and therefore shrinking and shortening, while the external wall with very little load to carry (at least initially) and exposed to the British weather was not shortening at the same rate.

Discussion — The porous LECA pellets were soaked just before the mixing of the concrete, to prevent them from absorbing water from the fresh mix and thus making it too stiff. They therefore constituted a reservoir of water, over and above that required for the hydration of the cement. This extra water meant that the LECA concrete needed more time to dry out and the 300 mm external walls would have a slower rate of drying out than the 250 mm internal walls even if they had the same environment on both faces.

Comment — These cracks were due to the changed properties of the wall concrete. A proper study of the properties of the materials along with a review of the design would have shown that the two-stage extrapolation from medium rise to high rise and from lean mix, brick rubble, unreinforced concrete to dense, albeit lightweight aggregate, concrete could not be sustained.

Chapter 8

Problems and Failures due to Poor Construction

8.1 FLAT SLAB CONSTRUCTION FOR HOTEL

For a short time in the early 1970s, the government provided loans for the construction of hotels. In order for a project to be eligible, the construction period had to be very tight. The workmanship of some of the hotels built then was shoddy. For one such hotel, the shoddy construction was not discovered until 20 years later when a major refurbishment was taking place.

Figure 8.1 shows the structural layout of a typical flat slab floor. The depth of the slab was 250 mm. The spans along the building were 7.2 m and across the building were 6.1 and 7.4 m. The top surface of the slab was very uneven and did not appear to have been levelled (by hand or power float). In some places, boot marks had been left. Cracks (generally not larger than 0.3 mm width) occurred on the upper surface radiating from the corners of the columns with one or two small cracks running tangentially. Large cracks (up to 1 mm width) appeared at some of the construction joints. The deflection of one of the slab bays of an upper floor was large—over 75 mm.

Several organisations became involved in assessing the situation. It was unfortunate that one of them concluded that the structure was unsafe and that one of the floors was in danger of imminent collapse. One bay of an upper floor in question was then set aside for instrumentation (both complicated and costly). After nearly a year of monitoring movement, the results showed no perceptible increase of deflection. However, during that period, additional steel brackets had been designed to prevent the slabs failing by punching shear, and they had already been fitted to the column–slab junctions of several floors. A further independent check was instigated and the following were its findings:

Excessive deflection — In order to understand how the excessive deflection had come about, it was necessary to check not only the design, but the in situ concrete strength, reinforcement details, and cover to the reinforcement as built. This check found that the only factor not as specified was the top cover to the reinforcement near the column supports. This was found to be on average 30 mm more than specified. This reduced the moment of

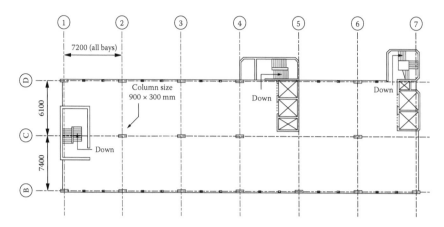

Figure 8.1 Structural layout of hotel floor.

resistance at the support. However, after reasonable moment redistribution was included in the calculations, there was sufficient overall moment capacity in the slab without requiring any reduction to the design safety factors.

This conclusion raised the question of why such large deflections had occurred. On close examination, it was noticed that the edge deflection of the slab along grid line B was also very large. In one of the bays, the skirting board between two edge columns had been made in two equal lengths split in the middle (see Figure 8.2). Deflections of 15 to 20 mm occurred below each half of the skirting board. This represented an edge deflection up to 50 mm. Since the skirting board was attached to the wall, it was likely that it was fitted this way and that much of the deflection had taken place before construction of the wall. This was confirmed by finding that the bottom courses of the external wall had been laid on the sagging shape of the slab and the following courses adjusted so that they were level at the window sills above the floor.

The conclusion was that much of the deflection occurred during construction, some of which was due to the sagging of the formwork and some due to early removal of the formwork.

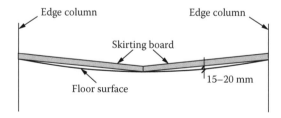

Figure 8.2 Skirting board deflection.

Punching shear — Punching shear failure is difficult to predict. Failure can occur with little warning. The presence of cracks in the slab, tangential to a circle around the column in the top surface of the slab, indicated that the possibility of a punching shear failure should be considered. The cover to the reinforcement was larger than specified in this area which meant that the crack size was magnified. However, the fact that the surface cracks were large did not necessarily mean impending shear failure.

A reliable method of predicting punching shear failure is by using the code design expression (*BS 8110*[1] or *BS EN 1992-1-1*[2]) for shear with the characteristic values for the concrete strength instead of the factored design values and the as-built information concerning the reinforcement (i.e., size, spacing and cover to the bars). An assessment of safety can be made by comparing the 'worst credible' loads with the resistance as calculated above. The calculations for this case showed that the worst credible loads could be carried with a sufficient safety factor.

Comment — It was unfortunate that the reason for the excessive deflection was not found earlier and that calculations were not made to check the punching shear capacity. Much costly remedial work could have been saved.

8.2 STEEL PILES SUPPORTING BLOCK OF FLATS

The pile cap supporting the structure for a multi-storey block of flats incorporated steel H piles. These should have been cut off close to the bottom of the pile cap but were taken up to within 100 mm of the top. The piles were coated with bitumen that had not been effectively removed from the protruding lengths that had to transfer the load from the pile caps through bond. The result was that as the building work continued, the pile caps started to settle with the piles acting as pistons. The concrete at the top of the pile caps failed in punching. The remedial work included digging under the pile caps and welding large collars around the piles to prevent further movement.

8.3 SHEAR CRACKS IN PRECAST T UNITS

Remedial work to the bearing of a precast column required local jacking of precast T floor units for several floors above to release their load. The jacks and props were taken right down the structure to the ground. At one floor, the props above and below did not line up. The eccentric load cracked the T unit at that floor. Figure 8.3 shows the opening up of 2 mm shear cracks.

The subcontractor proposed to repair the damaged T unit by drilling angled holes down the centre of the web and grouting in straight deformed reinforcing bars. He was sufficiently confident to agree to then test load

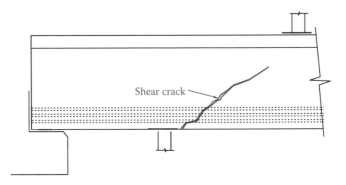

Figure 8.3 **Cracked T unit.**

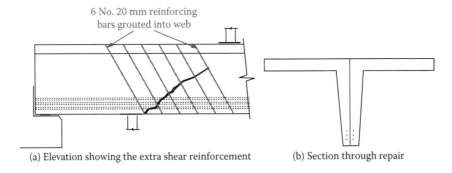

(a) Elevation showing the extra shear reinforcement   (b) Section through repair

Figure 8.4 **Repair of T unit.**

that unit with the full ultimate design load to check that it was serviceable. Figure 8.4 shows how the bars were arranged. The repair work required careful drilling to ensure that the prestressing tendons (shown dashed) were not damaged. The unit passed the extreme test load and permission was given for it to remain as part of the structure.

Comment — It was surprising that straight bars were capable of generating enough bond and providing sufficient shear resistance.

8.4 CANTILEVER BALCONIES TO BLOCK OF FLATS

Five years after construction of the flats, cracks appeared on the top surfaces of the cantilever balconies. Before carrying out repairs, investigation revealed that the reinforcement required near the top surface was placed half way down the section. Although the amount of cracking varied at each floor level, it was decided to rebuild the balconies at every floor level.

The engineer was curious because the fifth floor level showed no signs of cracking at all. When the balcony at that level was demolished, it was discovered that the reinforcement had been left out altogether!

The concrete had survived so well because its tension strength was sufficient to resist all the loads that the slab sustained. The reason for the cracking of the other balconies was that the reinforcement restrained the shrinkage of the concrete and thus created tension stresses in the concrete. Normally the reinforcement should be detailed with bars at small spacing. If the balconies had been constructed with the specified concrete cover, any cracking would have had a small width and been at a small pitch.

Comment — Although the upper balcony showed no signs of failure it was much more dangerous than those for the floors below. If subjected to a large impact load, it would have collapsed suddenly.

8.5 PRECAST CONCRETE TANK

A liquid storage tank was constructed with precast wall panels. The diameter and height of the tank were 12.3 and 7 m, respectively (see Figure 8.5). The vertical panels were held in place by unbonded prestressed tendons threaded through horizontal PVC ducts embedded in the concrete and fully encircling the tank at set levels throughout the height. The tank collapsed without warning within 2 years of construction. Failure was caused by a number of separate mistakes.

Water tightness — In order for the tank to be watertight, each precast vertical panel had to butt up to the adjacent panel evenly throughout the 7 m height. A rubber strip was inserted within the joint between each set of adjacent panels and incorporated holes through which the prestressing tendons were threaded. The prestressing was intended to create uniform compression throughout the height of the tank. However, to achieve

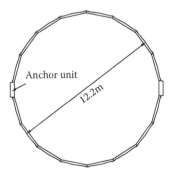

Figure 8.5 Precast elements of tank.

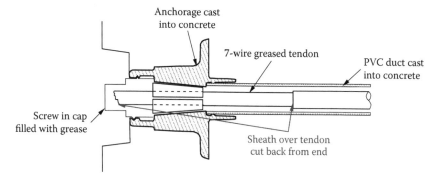

Figure 8.6 **Section through wall panel.**

Figure 8.7 **Detail of anchorage zone.**

watertight construction, the edges of the wall units had to be built with very small tolerances. The design of these precast panels (see Figure 8.6) allowed for several different diameter tanks. This meant that the angle of the edge formwork for the panel was adjustable and had to be set with extreme care for each diameter of tank. The water tests showed leaks. Several attempts were made to seal them before water tightness was achieved.

Prestressing ducts — During the water tests and possibly afterward, water penetrated the PVC prestressing ducts. The design assumed that even if water had penetrated the ducts, there would be sufficient protection of the prestressing tendons from the enclosing sheath and grease.

Sheathing and grease — The sheath and grease provided a continuous covering of the tendons up to the anchorage zone. In order to attach the prestressing jacks and insert the wedges, the tendons were stripped of the sheathing for some distance (see Figure 8.7).

The amount of sheathing required to be cut back depended on the extent of tendon stretch during prestressing. Inevitably after the operations of stressing and locking off, there remained a length of bare tendon. It would have been overly optimistic to assume that the grease cover to the stripped strands would be intact after threading them through the anchorage. Although possible, it would have been a difficult task to replace the grease around the bare tendon after stressing and unlikely to be completed successfully. Any water that penetrated the prestressing ducts would have reached this part of the tendon.

Grease — The grease used in this particular type of unbonded tendon (12.5 mm diameter Tyesa seven-wire strand manufactured in Spain) was found to emulsify when in contact with water. This allowed any water that penetrated the anchor zone to not only come into contact with the bare parts of the tendons but also to penetrate the sheathing.

Stress-corrosion cracking — The alloy steel of the prestressing tendons used in this structure had a microstructure susceptible to stress–corrosion cracking, and the stress in the tendons was greater than 50% of the yield strength. Moisture in contact with the tendons provided a corrosive environment. On examination after the collapse, it was found that stress–corrosion cracking had taken place in many parts of the unbonded tendons.

Comment — If similar strand manufactured in the UK had been used, the grease would not have emulsified. Nevertheless, it would still have been a difficult task to ensure that the bared ends of strands near the anchorages were adequately protected. Although it was less likely that stress–corrosion cracking would take place, there was still a significant risk. There is no reason to suppose that other tanks of this type did not leak when water tested as it was an exacting task to ensure that each vertical face of every panel matched up exactly with its partner. Hence it is likely that the prestressing ducts for many of such tanks would be flooded at some stage in their lives. In the opinion of the author, the vulnerability of such construction casts doubt on the viability and safety of such systems.

8.6 CAR PARK

In March 1997, a 120 tonne section of the roof of a car park collapsed onto the floor below (see Figure 8.8). This occurred at 3 a.m. when, fortunately, no people were in the structure. It was immediately clear from the debris that a punching shear failure had taken place.

Figure 8.8 Collapse of roof of car park. (From Jonathan Wood, http://www.hse.gov.uk/research/misc/pipersrowpt1.pdf.)

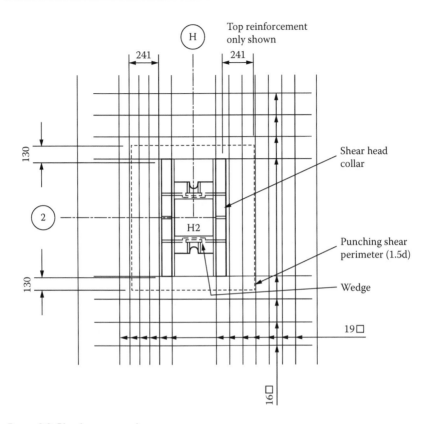

Figure 8.9 **Plan layout at column.**

The car park was constructed using the lift slab method. This involved casting the slabs one on top of another on the ground. Precast columns were positioned and then the slabs were jacked up the columns to the correct level. Final connection between the slab and the column was made via a steel collar in the slab and a steel insert in the column into which wedges were fixed. The steel collar supported the slab on angles that either formed a square or an H in plan. Figure 8.9 shows the typical H configuration used for internal columns. The column connection is very different from that found in a typical flat slab.

The 230 mm thick slab was constructed with concrete of highly variable quality. Areas of low quality concrete deteriorated, probably through freeze–thaw action. In some places, this deterioration occurred to a depth of 100 mm and had been repaired. The repair was poorly bonded to the parent material. This left a slab that was effectively split into two layers with the only connection being the longitudinal steel passing through the repair into the original concrete. Further deterioration of the original concrete, and in particular its bond strength to the top steel, reduced what composite action existed until failure occurred.

Comment — This form of construction had been used in many places in the UK during the 1970s and 1980s and has been a common form of construction in the U.S. It has provided reasonably robust structures. The very nature of the construction method focuses attention on the column–slab joint. In some situations, the structure has relied on the moment resistance of these joints, i.e., unbraced frames. In other situations, separate in situ core structures were built to handle the sway forces to which the complete structure may have been subjected. In some of the later examples, U bars were welded to the steel collars and embedded into the surrounding concrete.

8.7 CRACKING OF OFFSHORE PLATFORM DURING CONSTRUCTION

The substructure to this platform consisted of three concrete towers set on a cellular concrete structure that would sit on the sea bed in the final location (see Figure 8.10). It was constructed on shore in a dry dock.

During the operation of floating the platform out to the oil field and sinking it into the correct position, it was necessary to fill some of the cells with water (see Figure 8.11). This ensured that the platform floated at the right level as it was towed to the oil field. The cells and shafts were then filled in a controlled manner to allow the platform to settle on the sea bed into the correct position. The cells were completely filled with water during operation and none was used for oil storage.

Whilst the moving operation was taking place, the partially filled cells created differential pressures among the cells. It was important to ensure that the cells remained watertight during this operation. Arrangements

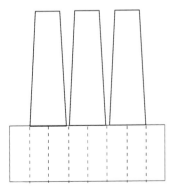

Figure 8.10 Elevation of offshore platform.

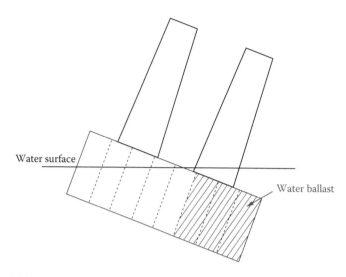

Figure 8.11 Water ballast pumped into some of the cells during sinking operation.

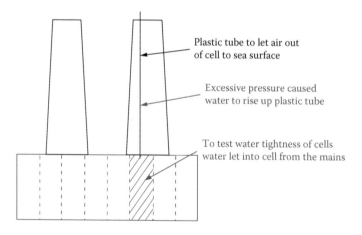

Figure 8.12 Water tightness test during construction.

were made during construction to test the water tightness of the cells. The intention was to fill each one of the cells to its top only. Each cell was constructed with a plastic tube that led to the top of the shafts to allow air to escape from the cell when it was filled during the final installation. The main water supply used to fill cells was capable of providing a pressure much more than that required, and the water level was allowed to rise up the plastic tube (see Figure 8.12).

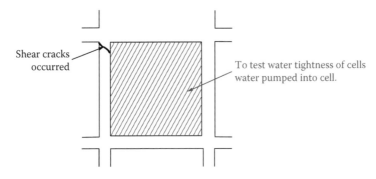

Shear cracks occurred

To test water tightness of cells water pumped into cell.

Figure 8.13 **Shear crack in wall.**

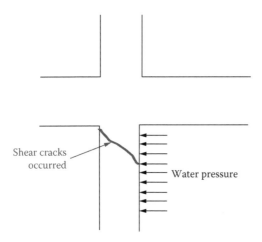

Shear cracks occurred

Water pressure

Figure 8.14 **Enlarged view of shear crack.**

Only a very tiny amount of water was needed to fill the vent pipe, but this vastly increased the hydrostatic pressure. The greatly increased water pressure in the cells exceeded the design pressure and caused one of the walls to crack (Figures 8.13 and 8.14). The repair work included casting reinforced concrete buttresses either side of the cracked wall and prestressing them to the wall with Macalloy bars (see Figure 8.15).

Comment — This failure could have been avoided if it had been realised that the testing arrangement could permit too high a pressure to be applied to the structure. The failure lay in the inadequate monitoring and control of the water pressure in the cell, not in the design of the structure. The overload was about three times the design pressure.

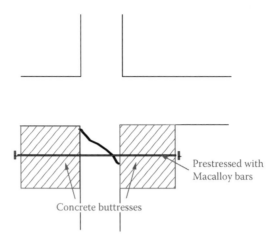

Figure 8.15 **Repair work.**

8.8 SPALLING OF LOAD BEARING MULLIONS

The facades of a tall office block were made of precast concrete sculptured panels (see Figure 8.16) and incorporated load bearing mullions. The horizontal panels and mullion joints were about 600 mm above the floor level. It had been specified that the panels be levelled by means of steel shims and then mortar beds be laid to a level just proud of the shims. The panels with the mullions attached would then be lowered down onto the beds with the intention that the load would be shared between the mortar and shims.

Sometime after construction, some of the mullions began to spall around the bearings. It was discovered that some of the mortar beds were missing. In place of these, the contractor had undertaken to dry pack the joints, but this had not been carried out thoroughly, probably because it was an impractical task. The mortar used was much softer than the steel shims and did not carry its share of the load that was largely transferred through the shims. The concentration of the load on such a small area of concrete in the mullions caused severe spalling (see Figures 8.17 and 8.18).

The repair work included inserting a set of steel columns at every floor (see Figure 8.19). To ensure the correct load was taken by each column, flat jacks were inserted at each floor level. Each jack was inflated with resin that was allowed to harden once the correct pressure was attained.

Comment — This joint detail might have worked had hardwood plywood been used for the shims instead of steel. The E value would have been close to that of the set mortar and, if pre-soaked, the shims would have shrunk as they dried, thus transferring the load to the mortar.

Figure 8.16 Office block with sculptured panel facades. (Courtesy of Poul Beckmann.)

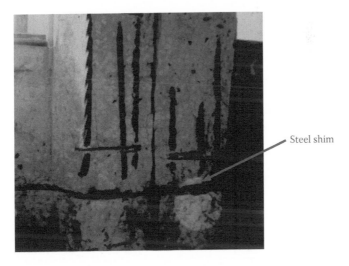

Steel shim

Figure 8.17 Spalling of mullions. (Courtesy of Poul Beckmann.)

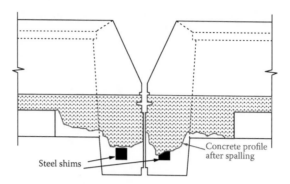

Figure 8.18 Plan section through mullion after spalling.

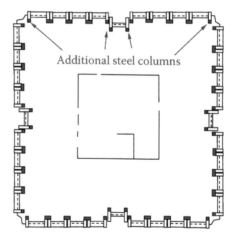

Figure 8.19 Layout of additional columns. (Courtesy of Alan Steele.)

Although this incident has been included in the chapter based on poor construction, the design detail for the joint between mullions was difficult, if not impossible, to construct as intended. It was critically important for the structural integrity of the facade and more thought should have been given to producing a more practical detail.

8.9 TWO-WAY SPANNING SLAB

A two-way spanning slab was designed in accordance with the design rules. One side was significantly longer than the other and this meant that the reinforcement provided for the short span was considerably greater than that for the long span.

When the drawing arrived on site, the engineer decided that the drawing contained a mistake and that the reinforcement in the long span should be the greater. He instructed the reinforcement layout to be swapped around and the slab was built that way. Later, when the full load was applied, the slab collapsed.

Comment — This failure could have easily been avoided if the site engineer had checked with the designer before making his decision.

8.10 CHIMNEY FLUE FOR COAL-FIRED POWER STATION

The wind shield of this chimney was 270 m high and 20 m in diameter. At low level, three 'portals' for the horizontal boiler flues were to connect with the three vertical flues. The flues were about 8 m in diameter and lined with firebricks. Their wall thickness was reduced above the portals by a taper on the outside surface. The reinforced concrete shafts of the wind shield and the flues were constructed simultaneously, using slip forming. The concrete was made with a high proportion of blast-furnace slag cement replacement; the Portland cement and the slag were not premixed, but supplied separately and stored in separate silos at the batching plant that had a single mixer.

The roof slab and one or two internal floor slabs were separated from the flue walls with Flexcel or a similar joint material to allow unrestrained temperature expansion. The firebrick linings to the flues were supported on corbels, and the casting and the laying of the brickwork followed the slip forming.

Above the roof slab at the top, the flues were also clad externally with brickwork. The bricklayers had just finished capping the last flue when a fracture lower down caused the top half of one of the flues to slide down, disintegrate, and fill the bottom of the wind shield with debris that spilled out of the portal for that flue. One worker was killed (see Figure 8.20).

It was still possible to enter the chimney space through one of the other portals, and the inside of one of the intact flues could be examined. A substantial horizontal crack was found on part of the wall surface; inserting a knife blade into the crack indicated that the fracture surface sloped upward. Subsequently, more cracks of a similar nature were found on the inside of the wind shield.

The fracture surfaces on the concrete fragments in the debris seemed more square and sharper than expected. This was later explained as a feature of sudden disintegration of high-strength concrete by impact, as opposed to the slow failures observed in laboratory tests. Cores were taken

Detail: Close up of debris

Figure 8.20 Debris from collapsed flue. (Courtesy of Alan Steele.)

at various locations. A number showed patches of blue colour on the freshly drilled surfaces, some distance from the wall faces. Hydrated slag is blue until it has been oxidized by exposure to the atmosphere. This therefore indicated that the cement and slag had not been properly intermixed. A number of cores, some of which were about 0.75 m long, were sawn into testable lengths and all showed adequate strengths. Other cores were taken across the horizontal cracks and showed clear separation.

An experienced slip forming expert was called in to give a second opinion. He explained that the horizontal cracks were lifting cracks that can occur when lifting is resumed after concreting has been interrupted. If some of the half-set concrete then sticks to the form, it is lifted with it, until the weight of the fresh concrete deposited on top forces it to drop back. As the concrete may not all drop neatly in one piece, cavities may result, obviously weakening the wall. Lifting cracks can be caused or aggravated by unevenness of formwork surfaces; they are usually hidden by the slurry rub-down that is customarily carried out from a finishing platform suspended some 2 m below a working platform.

Records and anecdotal evidence indicated a number of occurrences:

- A short distance above the foundation, a whole band had been cast with practically pure slag. This was discovered early, cut out, and re-concreted.
- Where the wall thickness was reduced above the flue portals, the whole of the formwork had to be lifted clear of the concrete to allow adjustment of the outside formwork to the smaller diameter. This had to be done twice—at the beginning and at the end of the taper. There was some congestion of reinforcement in this zone, some of it due to design, some of it due to inept steel fixing.

- Deliveries of materials were sometimes late, causing concreting to halt.
- The mixer broke down, causing concreting to halt.
- Supervision and inspection were inadequate.

The conclusion of how the collapse occurred was as follows:

- On occasion when concreting stopped, a severe lifting crack formed in the flue, extending about three quarters of the circumference. Because of its location some way up the shaft, the intact quadrant could carry the weight of the shaft above, albeit with a substantially reduced factor of safety.
- Solar heating caused the wind shield to bend. Theodolite measurements showed that the top of the wind shield on sunny days described an irregular horizontal orbit with a diameter in the order of 0.75 m.
- The flues had to follow this movement, which at certain times of the day increased the critical stress on the intact quadrant of the cracked flue. It also caused the lifting crack to open and close on a diurnal cycle. This caused the concrete to grind away just beyond the ends of the crack, gradually reducing the area of load carrying concrete until failure occurred.

Comment — More thorough supervision would have prevented this failure from occurring. The slip forming process requires careful monitoring at all times.

Chapter 9

Problems and Failures due to Poor Management

Many of the case studies in the other chapters may have been avoided with better management. The three cases reported in this chapter emphasise how important an engineering background is to the management of structural engineering projects.

9.1 COLUMN–SLAB JOINT

At times when so much work makes a project team over-committed, it is common to assign packages of work to other teams. For this particular job, a project team designed the slabs and handed over the design of the columns to another team. The reinforcement detailing was carried out by another group. All the design work should have been checked by the project team but in this case the separate parts of the detail design were passed to the detailing group without an overall check. It was common to detail the slabs and columns on separate drawings. Figure 9.1 shows the arrangement of the edge column–slab joint.

The reinforcement arrangement intended for this joint is shown in Figure 9.2. Construction reached the second floor when a young graduate visited the building to gain some site experience. He reported back his concern about the reinforcement layout along the edge of the slab that was about to be concreted. Figure 9.3 shows what he saw. There was no mechanical link between the slab and the column reinforcement! It was immediately realised that not only had this occurred at all edges of the slab on the second floor, but also on the first floor that was completed several weeks earlier.

Immediate remedial work was put into action. Temporary supports were put in place for both floors and then holes were drilled through each joint and long bolts grouted in place (see Figure 9.4).

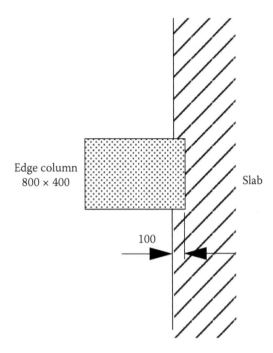

Figure 9.1 Plan layout of edge column–slab joint.

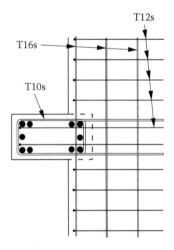

Figure 9.2 Intended layout of reinforcement.

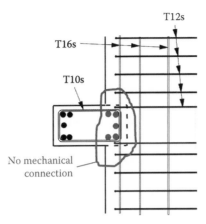

Figure 9.3 **Actual layout of reinforcement.**

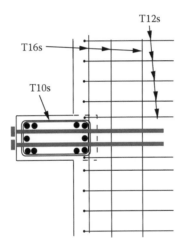

Figure 9.4 **Remedial work.**

Comment — One of the reasons this situation occurred was that no check had been made in the design office to ensure that a mechanical link could be made. In fact, when the cover and the actual size of bars were taken into account, it was clear that the reinforcement could not fit as intended. The fabricator of the reinforcement fitted the bars as close to the correct position as possible without thinking that a mechanical link was essential.

Nevertheless, the underlying cause of this mistake was in failure to manage the design.

9.2 PLACING OF PRECAST UNITS

A spine beam carrying precast planks lost its bearing because a labourer trying to jack one of the final planks into position actually levered out

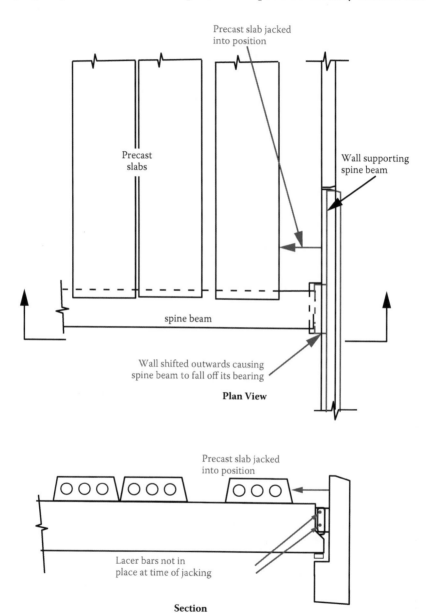

Figure 9.5 Adjusting positions of precast planks.

the wall panel supporting the end of the spine beam (see Figure 9.5). This caused the spine beam to lose its bearing and led to the collapse of the floor. The connection between the spine beam and the wall panel had been designed to have two overlapping U bars locked together with lacing bars threaded through the space between them. These lacer bars had not been inserted at the time of erecting and laying of the floor elements. If they had been in place, they would have prevented the wall panel from moving away from the spine beam.

Comment — This is an example in which management should have had more control on how the erection and placing of precast units took place and, more importantly, ensured that the lacer bars at the ends of the spine beam were in place before the erection of floor units took place.

9.3 WEAK AGGREGATE CONCRETE IN CHIMNEY

A tall chimney with a single flue shaft for a coal-to-petrol synthesizing plant was being slip-formed. Half-way up, the 3-day cylinder tests results suddenly showed a serious dip in the strengths. The samples had been taken from concrete placed between midnight and the following morning. Subsequent test results were satisfactory. After some probing and questioning, the following scenario emerged.

There were two batching plants on the site, each with its own day-work stockpile. One was supplied with locally available, but somewhat inferior, aggregate for use in low-grade concrete for plant foundation blocks. The other had to obtain 'good' aggregate for the chimney from some distance away. The batching plants were at opposite corners of the site.

On the night in question, the midnight aggregate lorry did not appear on time. The foreman in charge of slip-forming found himself facing an unplanned stop to the sliding that would have caused a calamity. Apart from the resulting delay, it could easily have caused a problem when restarting the slip-forming, for which he could have been blamed. Not aware of the different qualities of the aggregates, his solution was to get a couple of dumper trucks to fetch aggregate from the other batching plant that was shut down for the night. The sliding could then continue.

However, this left the other batching plant short of aggregate. When the lorry with the 'good' aggregate arrived in the small hours, the sliding foreman directed it to unload at the low-grade batching plant to avoid a daybreak dispute with that foreman. Normal service was then resumed.

By the time all this came to light, a substantial height of good concrete had been placed on top of the low-grade aggregate belt. To avoid demolishing some 20 m of shaft, it was proposed to construct a "collar" or "sleeve" of good concrete or gunite around the weak zone. Composite action could be provided by drilled-in anchor bolts acting as shear connectors. This would necessitate a planned interruption of the sliding to enable a working platform to be suspended from the slip form, so the remedial work could be carried out, but this was accepted as a necessary delay.

Comment — Careful management is essential for the slip forming process. There should have been procedures in place that ensured consistent supplies of the correct aggregate. The stockpile should not have been allowed to fall to such a level that the slip forming production line was waiting for a lorry to arrive.

Chapter 10

Problems and Failures due to Poor Construction Planning

Planning of any construction work requires the input of engineering thought and this includes the time needed to consider 'what ifs?'. Hand sketches can be very useful to explain what should or could happen.

10.1 POWER STATION ON RIVER THAMES

A power station was constructed on the north bank of the Thames in the early 1960s. Originally it was to be coal fired to produce 1500 MW. The foundations of the power station sat on 20,000 reinforced concrete piles and a special casting yard was set up on site to produce them (see Figure 10.1).

The piles were 430 mm square and 18 m long. Several pile rigs were set up with diesel-driven hammers (see Figure 10.2). A pile was hoisted into position and then given a tap by the hammer to get the point of the pile through the top crust of the marshland. Then under its own weight the pile dropped 15 m through the mud. Each pile was then driven into the gravel until a specified set had been reached.

Piling commenced from the edge of the site closest to the river and continued inland for a distance of over 250 m. Piles were placed at 1.5 m centres (on average). The exact positions of piles shown on the drawings related to the type of foundation (turbine, boiler, culvert, or miscellaneous). The time period for this part of the project was about 18 months. Excavation for the foundations, also starting from the river end of the site, commenced 6 months after the start of piling. The depth of excavation varied from 1 to 3 m. This exposed the piles that were then cut down to the level of the concrete blinding. Concreting of the foundations then commenced, starting from the same end of the site as the piling and excavation.

A year after the start of piling, when concreting of the foundations progressed about a third of the way along the site, it was discovered that the tops of the piles that were still exposed were moving. *Measurements showed that the movement was up to 1.5m!*

Figure 10.1 **Pile casting yard.**

Figure 10.2 **Pile driver.**

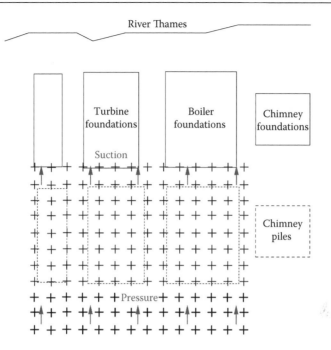

Figure 10.3 **Diagram showing cause of movement.**

The first major concern was to check whether the piles were still intact. One pile was extracted. It came out straight and appeared undamaged. On close examination, fine cracks could be seen throughout its length. This indicated that it had bent evenly over its full length.

It took some weeks to fully understand what had happened. Figure 10.3 shows a plan of the site and attempts to demonstrate the situation. The steady addition of piles into the ground built up ground pressure. This pressure was applied increasingly from one direction. At the same time, the excavation for the foundations of the structure closest to the river released any ground pressure that built up. The combination of these two major actions caused the movement.

The resulting remedial work included

(1) adding an additional 600 vertical piles to compensate the reduction in vertical capacity of the existing piles
(2) adding 200 more raked piles to compensate for the horizontal force component caused by the bent piles.

Large amounts of remedial and extra work were required because of the movement of all these piles. For example, the existing piles no longer followed the plan layout for the eight inlet and outlet culverts that wound their way through the site to bring cooling water to the condensers and return it to the River Thames. On-site decisions to make changes to the design had to be made each day. It became very tricky trying to fit the extra piles, especially for the 170 m chimney pile caps. The original design included a careful arrangement of raked piles which had been driven all pointing towards the centre of the cap. They had all tipped over, making a complete tangle. As the extra piles were driven down they kept hitting existing piles. This meant that they had to be repositioned and angled to obtain a clear route.

Comment — The programme for the contracts on this project did not foresee the problems caused as the work progressed from one end of the site to the other. In previous similar projects, a significant delay between piling and the start of excavation allowed enough time for much of the soil pressure to dissipate. The need to reduce time and costs on this project was not balanced by careful consideration of the consequence.

Where piling layout drawings are made for large sites, each pile is indicated by a cross on the drawing, so the volumetric effect of each pile and the entire group is not immediately apparent. To keep a tight programme, one possible solution might have been to start the piling from both ends of the site.

10.2 TOWER BLOCK

An underground car park was to be added to one of several residential tower blocks. Excavation commenced on the south side of the block and the excavated soil was piled on the north side to a height of 10 m (see Figure 10.4). A heavy rainfall occurred some time after the excavation finished and it increased the lateral ground pressure on the piles. The building began to move sideways toward the excavation. This caused the piles to fail as their shear resistance was exceeded (see Figure 10.5). The building became unstable as it moved toward the excavation and eventually fell over (see Figure 10.6).

It was very fortunate that the other tower blocks were not close enough to cause a domino effect. Figures 10.7 and 10.8 show how close this tower block was from its neighbours. Figure 10.8 shows a close-up view of the piles that failed in shear. The piles were made of circular hollow unreinforced concrete. They were constructed with short lengths at the top with solid reinforced sections (see Figure 10.9).

Comment — There appears to have been no engineering planning to prevent the combination of excavation and piling up of the soil on opposite

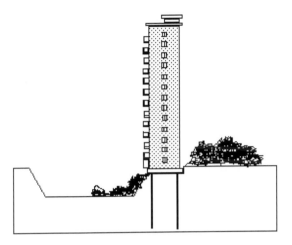

Figure 10.4 Diagrammatic view after excavation.

Figure 10.5 Diagrammatic view of pile shear failure.

sides of the building. Some hand sketches showing the intentions of the plan would have helped identify the likely problem.

It was probably reasonable for the piles to be designed for compression only since the loading to which the foundations were subject was beyond the scope of the design.

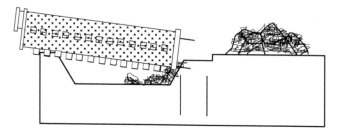

Figure 10.6 Diagrammatic view of collapse.

Figure 10.7 View of underside of collapsed building. (Courtesy of Gillian Whittle; redrawn
from Reuters, http://www.telegraph.co.uk/news/worldnews/asia/china/5685963/
Nine-held-over-Shanghai-building-collapse.html.)

Figure 10.8 View of piles that failed in shear. (Courtesy of Gillian Whittle; redrawn from Reuters,   http://www.telegraph.co.uk/news/worldnews/asia/china/5685963/ Nine-held-over-Shanghai-building-collapse.html.)

Figure 10.9 Unreinforced piles with short reinforced length at tops. (Courtesy of Gillian Whittle; redrawn from Reuters, http://www.telegraph.co.uk/news/world-news/asia/china/5685963/Nine-held-over-Shanghai-building-collapse.html.)

Chapter 11

Problems and Failures due to Deliberate Malpractice

11.1 FLOOR WITH EXCESSIVE DEFLECTION

The building in question was a telephone exchange built in the mid-1970s, 10 years earlier. Figure 11.1 shows a plan and section of a typical end bay. The slab had been designed as single way spanning between two shallow haunched beams. The design included a span of 9 m with a slab only 250 mm thick which many engineers would consider too thin.

Ten years after the building had been completed, the operators complained that the deflection was still increasing and had caused some of the switch gear to become faulty. The designers asked for a second opinion on the design of the slab. The calculations and drawings were checked and no major flaws were found. It was conceivable that creep and shrinkage effects were still increasing.

The designers also asked the local university to run an independent check using its finite element modelling package. The subsequent report concluded that there was some major overstressing that could lead to a shear failure. The finite element analysis assumed elastic behaviour that led to high stresses at the supports. This was emphasised by the shallow supporting beams that allowed the slab to behave more like a flat slab. However, if a reasonable amount of moment redistribution was assumed, although some cracking of the concrete could be expected, the design was safe. Nevertheless it was difficult to convince the designer that this was so and a visit to the site was arranged.

The visit to site included the inspection of the slab close to a column. The screed had been removed to expose the top surface of the structural slab. On close inspection, the reinforcement appeared to be badly rusted and breaking through the top surface. As a crude check of the hardness of the concrete surface, it was scratched with a penknife. Quite unexpectedly, the blade of the knife penetrated into the concrete surface right up to the hilt! A further check of the soffit of the slab gave a similar result.

An additional interesting feature of the soffit was the presence of a number of shallow disc shaped ('flying saucer') pieces of concrete (150 mm

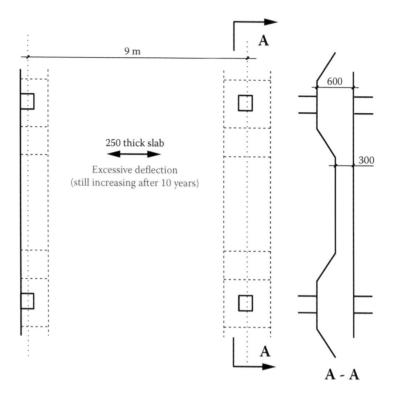

Figure 11.1 Plan and section of typical end bay.

diameter) that were separating from the surface. One such piece came away as it was being examined. Although the slab had been designed to span one way, the supporting beam was sufficiently flexible for the slab to behave more like a flat slab. The 'flying saucers' appeared in the compression areas of the soffit and were considered to be the effects of spalling. It was clear that the slab in question required immediate additional support and the rest of the building required core testing.

After cores had been taken throughout the building, it was discovered that the concrete cube strength that should have been 25 MPa was on average only 5 MPa. The subcontractor had deliberately reduced the cement content in the specified mix. Major remedial work followed.

Comment — It is very surprising that the building structure remained intact for so long and demonstrates that, if there is a possible force path to hold a structure together, the structure will find it. The client was relieved to learn the real cause of the problems. It is remarkable that the telephone exchange remained in operation throughout the period of remedial work.

11.2 PILES FOR LARGE STRUCTURE

Over thirty reinforced concrete piles (3 m diameter) were required to support a complex structure including a 3 m thick transfer slab and several tower blocks above. When the construction of a tower block above reached several storeys, it was discovered that the piles were less than half the required length.

Many ideas to resolve the situation were considered, but in the end the chosen solution was to provide a set of nine steel H section piles around the existing piles and drive them to the required depth. The H piles were then connected to the existing structure through new pile caps (one per original pile). To carry out the remedial work, one of the intermediate basement floors had to be removed to allow the piling rigs to operate with sufficient head room. Access into the basement was gained by cutting a large hole through the existing perimeter retaining wall.

Comment — The pile shortening was a deliberate act of the subcontractor that resulted in a long delay to the completion of the project at enormous extra cost.

11.3 IN SITU COLUMNS SUPPORTING PRECAST BUILDING

This building was constructed with precast elements above ground. Below ground, the foundations, columns, and beams were constructed in situ (see Figure 11.2). Construction had reached an advanced stage when cracks appeared in the in situ columns just below the connections with the precast columns.

The construction of the in situ columns should have proceeded as shown in Figure 11.3. The central circular hollow section (CHS) dowel was cast into the top of the in situ column ready to receive the precast column.

In order to check that the concrete in the box-out was in place as intended, holes were drilled through each face. It was discovered that the polystyrene had been left in and only a thin layer of concrete had been placed at the top (see Figure 11.4).

The load in the precast column from seven floors above transferred to the outer rim of the in situ column. This load was intended to be taken by the whole section of the in situ columns. However, the thin layer of concrete at the top of the box-out and the polystyrene below were quite incapable of taking any significant load. The outer shell of in situ concrete that had to take the load became overstressed and consequently cracked, providing the first sign of imminent failure (see Figure 11.5).

In order to repair the tops of the in situ columns, the load from the precast building had to be removed. This was achieved by providing props and

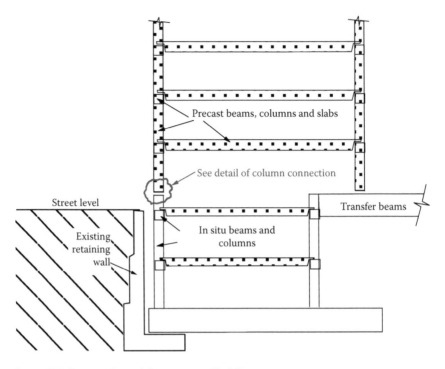

Figure 11.2 **Section through lower part of building.**

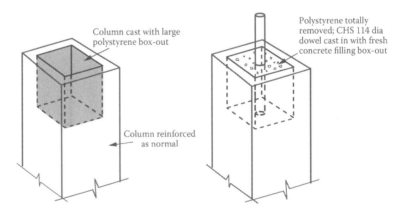

Figure 11.3 **Intended construction of top of in situ column.**

jacks close to the existing precast columns at each floor level and creating a new load path to the ground. This released the load on the in situ columns below and allowed the required remedial work—reconstruction of the tops of the in situ columns—to take place.

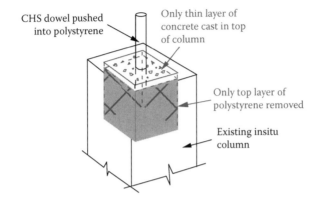

CHS dowel pushed
into polystyrene

Only thin layer of
concrete cast in top
of column

Only top layer of
polystyrene removed

Existing insitu
column

Figure 11.4 **Actual construction of top of in situ column.**

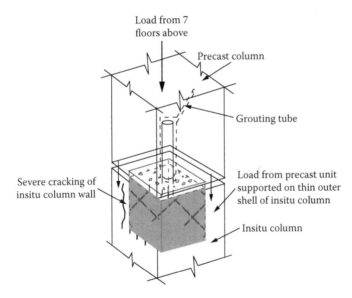

Load from 7
floors above

Precast column

Grouting tube

Severe cracking of
insitu column wall

Load from precast unit
supported on thin outer
shell of insitu column

Insitu column

Figure 11.5 **Cracking of outer rim of in situ column.**

Comment — Although the act that caused the problem was not intended to cause collapse of the building, it was, in the author's opinion, a deliberate act to leave most of the polystyrene box-out. This was done to save time and effort for the contractor. It was extremely fortunate that the fault was found before collapse of the building occurred.

Chapter 12

Problems Arising from the Procurement Process

12.1 EFFECTS OF DIFFERENT FORMS OF CONTRACTS

Typically the form of procurement lies between the two extremes:

- Traditional architect-led contracts (e.g., one off)
- Design–build contractor-led contracts (e.g., system build)

The former provides the client and architect with the greatest freedom of layout and form. The latter can lead to the most economic and fastest completion time. Both rely on the integrity and care of the whole team.

There are significant differences in some aspects of the design and detailing between these two forms of contracts. The reasons are not difficult to understand but they have a marked effect on the way the information is produced. Where a project is architect led, the contractor that will be awarded the work is not known at the time of the design. This means that the design and detailing should be carried out without involving proprietary systems unless they are required by the design. This is because each contractor is likely to favour a particular proprietary system that is not preferred by others. The detailing should include only reinforcement required by the design and not be controlled by possible site requirements (such as health and safety). This can be best explained through examples.

Example 1 — One method that contractors use to resolve the danger that people will trip over slab reinforcement before and during placing of concrete is to place top reinforcing bars at small centres (say, 200 mm or less). This is often not a design requirement. In fact, very often there is no design need for top reinforcement in the middle area of a slab.

Hence, if the drawings show only the reinforcement required for the design, the contractor will have to decide whether to propose extra reinforcement or provide some other method of ensuring safe passage for people (such as providing temporary planks and walkways). The client is interested in the cheapest method, not necessarily the most convenient one for the contractor. All too often, the extra reinforcement along with the extra cost is assumed to be a requirement of design when there may be cheaper ways of achieving the required safety.

Example 2 — Another common example concerns the protection of column reinforcement starter bars. Health and safety regulations have required the ends of the starter bars to be bent into hooks. This is costly to achieve and a cheaper alternative would be to place an open plastic box over the top of each group of column bars. The boxes could be reused for each floor.

Example 3 — A significant difference between the two forms of contracts concerns the positions of construction joints. For building structures, good practice ensures that reinforcement is lapped at positions of low stress. The lap and anchorage lengths used by detailers are often set by a simple "rule of thumb." Often 40× bar diameter is used. This is considered by most engineers to be a safe approach and normally no further checks are made.

However, some contractors realised that this is a point where savings can be made and for design–build contracts, they instruct the detailer to use 35× bar diameter. This would probably be acceptable for a traditional type of contract where the specification demands that construction joints be positioned away from highly stressed areas. However, for design–build contracts, a contractor often decides to position construction joints at positions convenient for construction—where reinforcement may be highly stressed. The required lap length in such a position could be as high as 60× bar diameter.

This causes a dilemma for a detailer. Should he or she use 60× bar diameter anchorage and lap lengths for design–build contracts to ensure safety in all situations?

12.2 WORKMANSHIP

The quality of workmanship is likely to remain the greatest issue within the industry for the foreseeable future and will rely on the intent and enthusiasm of the managers to improve standards. It is not just a question of ticking boxes on a form. It is about ensuring that the engineering of each part of a job is understood and carried out correctly. It requires individuals to take on the ownership and responsibility for the execution of good quality work. Workmanship quality is strongly affected by:

- An intelligent workforce and requirement for checking
- Achieving individual satisfaction and pride in work
- Utilizing sensible procedures to avoid mistakes

One criticism levelled against the traditional form of contract is that many designs are too conservative and as a result concrete construction is less economical than construction using other materials. It is certainly true that many designers err on the conservative side when sizing elements. This has come about from past experience when it became common for architect clients to make changes late on in the job, leading to increased loading or changes in structure (moving column positions, adding large holes, etc.).

During the construction phase, a contractor may see ways to reduce his costs and often this is possible because of the conservatism of the design. For a standard form of structure (e.g., office or residential block), this conservatism can be considered inefficient and wasteful.

At the other extreme, for some design–build contracts, the workmanship is considered to be of low quality. Occasionally methods adopted for a particular type of construction (e.g., hybrid car parks) are extrapolated for longer spans and are ill thought out, sometimes producing an unsafe structure.

It is not uncommon for a design–build contract to use hybrid concrete construction (in situ and precast concrete). This may well lead to the design of the individual elements by designers working for different companies. *In such situations it is essential that there should be a single responsibility of one engineer for the stability of the structure, and the compatibility of the design and details of the parts and components, even where some or all of the design, including the details of those parts and components, are not carried out by this engineer.*

The lack of systematic or third party checking has led to poor and in some cases unsafe construction (e.g., ungrouted prestressing ducts). Third party checking should be carried out where considered necessary by the engineer and, in the author's opinion, should be incorporated more frequently in all forms of contract.

12.3 CHECKING CONSTRUCTION

On one major project, in situ post-tensioning was required for the floor slabs. An independent resident engineer was not included in the contract and it was assumed that the contractor would provide sufficient supervision of the work.

After construction, water was discovered dripping from the soffit of the slab. Investigation revealed that the prestressing duct just above the leak had not been grouted. A further investigation required that all the ducts in the building had to be checked and this revealed that many other ducts had not been grouted. The possibility of stress corrosion and failure of prestressing tendons can be greatly increased by the presence of water as was the case with these ungrouted ducts. The remedial work to resolve this mistake was costly and time consuming.

Comment — In a prestressed floor, the amount of reinforcement is negligible compared to a floor with reinforcement only. The spacing of tendons can be up to 2 or 3 m. If one tendon were to fail, up to 6 m width of slab would remain unsupported. The risk of collapse of the slab and progressive collapse of the whole structure would be significant. There is a growing concern amongst many engineers that the quality of design and construction has declined as a result of the increasing pressure to cut the time and cost of projects.

It has been reported to CARES[13] that a survey has shown that a number of ducts, approaching 1% on average, have been either partially or completely ungrouted, thereby potentially affecting the integrity of a number of buildings in the long term. This is an average figure, and it is likely that the average has been exceeded greatly in a number of structures.

In many situations, the checking of construction work has reduced to unacceptable levels. There must be a balance between rigorous procedures and the amount of checking required. The present trend is to increase the requirements within the specifications. The author is not convinced that this is sufficient or the best way to improve the quality of work. Having a third party check for critical parts of a structure would appear to be a sensible solution.

Chapter 13

Contributions of Research and Development toward Avoidance of Failures

13.1 LINKS BETWEEN PRACTICE AND RESEARCH

The need for research and development continues to increase as materials, designs, and construction methods are refined. It is very important that good communication links are maintained between engineering practice and research institutions. This is often best achieved through personal contacts. It is unfortunate that such links are becoming less common in the UK, probably due to reductions in available funds. The following examples describe how the author has been involved in such links.

13.2 FLAT SLAB BEHAVIOUR

In the late 1970s, flat slab construction was becoming popular as it resulted in thinner slabs. It also allowed simple and quick means of construction. However, designers were looking for more tools to help them analyse such structures and demanding more information about their strength and performance. The author became involved in a research project to test flat slabs at the Polytechnic of Central London. This resulted from questions about the behaviour and strength of this form of construction. Dr. Paul Regan set up the tests (see Figure 13.1) and produced a ground breaking research report for CIRIA (*Report 89: Behaviour of reinforced concrete flat slabs*[14]).

One important area of doubt had been punching shear behaviour. Regan's tests provided valuable information on this subject, taking into account the effect of moment transfer between slab and column. This information has since been developed and included in both the UK and European codes of practice.

Figure 13.1 Flat slab tests carried out at Polytechnic of Central London. (Courtesy of Jonathan Wood.)

13.3 SPAN AND EFFECTIVE DEPTH RATIOS FOR SLABS

BS 8110[1] and previous UK codes of practice provided clauses on span and effective depth. Modification factors were provided for tension and compression reinforcement. In 1994, during the drafting of the Concrete Society's *Technical Report 49: Design guide for high strength concrete*,[15] it was realised that some concrete properties improved with strength. These included an increased modulus of elasticity, increase of cracking moment, and reductions in shrinkage and creep.

Professor Beeby, who provided the supporting information for the existing code clauses, was asked whether it was appropriate to make an adjustment to the span and/or effective value for the strength of concrete. After further research, he provided an expression to include this variable. Figure 13.2 shows a plot of the expression that was adopted in the technical report and later, in 2004, was also included within *Eurocode 2*.[2]

13.4 BEAM AND COLUMN JOINTS

Tests were carried out at Durham University to determine the stresses in both the concrete and the reinforcement as the load was increased to failure. Some of the reinforcing bars had special treatment to insert a large number of strain gauges into slots through the centres of the bars. Each bar consisted of two separate bars milled longitudinally so that cross-sections

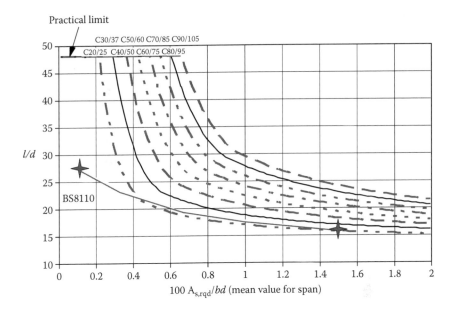

Figure 13.2 **Effect of concrete strength on span and effective depth ratios.**

were halved. A slot was cut down the centre of each of bar and the strain gauges placed along one bar at varying pitches (as little as 50 mm) and glued into position. The two halves were then joined to simulate a single complete bar. All the wires from the strain gauges were led out of the ends of the bars to a control box. Figure 13.3 shows one of the tests at the point where the joint failed.

Different arrangements of reinforcement were detailed and tested. One important result showed that if the beam top bars were bent upward in the column, with gravity loading on the beam, the joint resistance was drastically reduced compared with the bars bent down. The reason is that the high compression force created within the joint was not resisted by the enclosing bar as shown in Figure 13.3a.

13.5 TENSION STIFFENING OF CONCRETE

The importance of time effects of tension stiffening of concrete became apparent after testing was carried out at Leeds and Durham Universities. This work was undertaken as a result of some unanswered questions from a Brite Euram project on high strength concrete carried out in 1995. As part of that project Taywood Engineering built a full-scale thin flat slab (see Figures 13.4 and 13.5).

(a) Severe cracking   (b) Failure

Figure 13.3 Loading column joint to failure. (Courtesy of Richard Scott.)

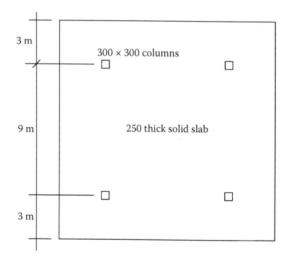

Figure 13.4 Layout of Taywood test flat slab.

The behaviour of the slab was carefully monitored during its early life and the author compared the measurements with the results from a finite element model. The results corresponded very closely up to and just after the full test load had been applied. It came as a surprise to discover that 2 weeks after unloading the slab, it deflected more than the model had predicted. Dr. Robert Vollum of Imperial College set up an independent model but the results did not explain the differences.

Figure 13.5 Construction of test flat slab. (Courtesy of Richard Scott.)

The solution to this problem came about as a result of a chance meeting with Prof. Andrew Beeby two years later. He suggested that he and Dr. Richard Scott carry out tests at Leeds and Durham Universities to examine the concrete tension stiffening effects in more detail. The project was an unmitigated success and resolved the problem with respect to the Taywood Engineering slab. It revealed that what had been considered short term effects of tension stiffening lasted much less time than assumed and that the long term effects took over after a few days. Figure 13.6 shows how tension stiffening affects the behaviour of a concrete element.

The knock-on effect of that project has been a significant change in the application of the British and European codes. Each of these codes stated that tension stiffening is halved in the long term. Until recently, *long term*

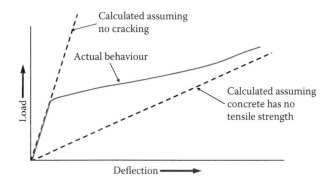

Figure 13.6 Effect of concrete tension stiffening on deflection of slabs.

was assumed to be several years after loading instead of the actual few days. This means that for design purposes the long term value should almost always be used.

This research has provided an advance in the understanding of the behaviour of reinforced concrete when it cracks. The report on the project has been converted to Concrete Society *Technical Report 59: Influence of tension stiffening on deflection of reinforced concrete structures.*[16]

References

1. British Standards Institution. *BS 8110-1 The structural use of concrete. Part 1: Code of practice for design and construction*, London, 1997.
2. British Standards Institution. *BS EN 1992-1-1: Eurocode 2, Design of concrete structures. General rules and rules for buildings*, London, 2004.
3. Construction Industry Research and Information Association. *Report 91: Early-age thermal crack control in concrete*, 1992. Replaced by publication C660 in 2007.
4. Whittle R. and Taylor H. *Design of hybrid concrete buildings*, Concrete Centre. 2009.
5. Concrete Society. *Technical Report 67: Movement, restraint and cracking in concrete structures*, 2008.
6. Concrete Society. *Technical Report 64: Guide to the design of reinforced concrete flat slabs*, 2007.
7. British Standards Institution. *BS EN 1994-1-1: Eurocode 4, Design of composite steel and concrete structures. General rules and rules for buildings*, London, 2004.
8. British Standards Institution. *CP 114: The structural use of reinforced concrete in buildings*, London, 1957.
9. Rowe R. E., Cranston W. B., and Best B. C. New concepts in the design of structural concrete. *Structural Engineer*, 43, 399–403, 1965.
10. British Standards Institution. *CP 110: The structural use of concrete*, London, 1972.
11. Department for Communities and Local Government. *Building regulations, 5th amendment*, HMSO, London, 1970.
12. British Standards Institution. *CP 116: The structural use of precast concrete*, London, 1965 (revised 1970).
13. United Kingdom Certification Authority for Reinforcing Steels (CARES).
14. Construction Industry Research and Information Association (CIRIA). *Report 89: Behaviour of reinforced concrete flat slabs*, 1981.
15. Concrete Society. *Technical Report 49: Design guide for high strength concrete*, 1998.
16. Concrete Society. *Technical Report 59: Influence of tension stiffening on deflection of reinforced concrete structures*, 2004.

Index

Printed and bound by CPI Group (UK) Ltd, Croydon, CR0 4YY

18/10/2024

01776245-0005